Samuel Hubbard Scudder

A classed and annotated bibliography of fossil insects

Samuel Hubbard Scudder

A classed and annotated bibliography of fossil insects

ISBN/EAN: 9783742830128

Manufactured in Europe, USA, Canada, Australia, Japa

Cover: Foto ©Klaus-Uwe Gerhardt /pixelio.de

Manufactured and distributed by brebook publishing software
(www.brebook.com)

Samuel Hubbard Scudder

A classed and annotated bibliography of fossil insects

BULLETIN

OF THE

UNITED STATES

GEOLOGICAL SURVEY

No. 69

WASHINGTON
GOVERNMENT PRINTING OFFICE
1890

UNITED STATES GEOLOGICAL SURVEY

J. W. POWELL, DIRECTOR

A

CLASSED AND ANNOTATED

BIBLIOGRAPHY OF FOSSIL INSECTS

BY

SAMUEL HUBBARD SCUDDER

WASHINGTON

GOVERNMENT PRINTING OFFICE

1890

CONTENTS.

5

PREFACE.

The present work is an extension to date of a bibliography published in 1882. It has, however, been altered in a few details, and, besides being fuller, differs from that in being a classed list, the works and essays which cover the entire field (which embraces not only insects proper, but also myriapods and arachnids) being placed first, followed by the more special memoirs grouped first by times, next by classes, orders, etc., the classification employed in my Systematic Review of Fossil Insects, being used as a convenient basis. This will also form the basis of the Index to Known Fossil Insects, forming a later complementary bulletin. The occasion for the publication of both of these at this time is the completion of the first extended account of the American Tertiary insects given in Vol. XIII of the Hayden series of geological reports, by which the numbers of the European and American insects bear for the first time some sort of proper relation to each other, at least in the lower groups. This makes an immediate "account of stock," to employ a commercial term, desirable.

The points wherein the present biliography differs from its predecessor, besides its classification and its inclusion of later material, are the somewhat fuller and otherwise altered notes, and the discontinuance of references to later treatises on the relations which Limulus and its allies bear to the arachnids. For the same reasons as before, viz, that the study of the Merostomata is ordinarily confined to a different group of paleontologists from those who are engaged on fossil insects, and because it would unduly, and, in my opinion, unwisely, extend the scope and special purpose of this bibliography, no attention is given to this side of the subject beyond the earlier and largely controversial papers; though it is not intended that this action shall be in any sense an expression of opinion, for this, not having specially studied the subjects, I am not prepared to give.

I have thought it best to retain the works upon amber given in the earlier bibliography, even when they make only the broadest allusion to insect inclusa, since the fossil insects of the European amber form such a predominating element in the tertiaries of the Old World, and I have even added a few later entries. Perhaps not one-half the works or papers concerning amber referred to in bibliographies are procurable in this country, and of those seen comparatively few contain references to insect inclosures. The scattered allusions to amber insects, taken from older authors and published later than the seventeenth century, have in general been left unnoticed as wholly valueless and uninteresting; but all others, where possible, have been introduced into the list in a more or less perfect form. The best notice of the early literature of amber will be found in Bochmer's Bibl. Script. Hist. Nat., 4, i, 468–477. 8°. Lipsiæ. 1788.

The Index of Authors at the end contains over five hundred names.

BIBLIOGRAPHY OF FOSSIL INSECTS.

I.—GENERAL AND MISCELLANEOUS.

Agassiz, Jean Louis Rodolphe. The primitive diversity and number of animals in geological times. 8°. [New Haven.] 1854. pp. 16. (Am. journ. sci., (2), 17 : 309-324.) 8°. New Haven. 1854.

Devotes a paragraph on p. 8 (316) to insects, and prophesies their discovery anterior to the Carboniferous period.

—— See also **Buckland, W.**

Aldrovandus, Ulysses. De animalibvs insectis libri septem cvm singvlorvm iconibvs ad vivvm expressis. f°. Bononiæ. 1638. t. p., pp. (8), 767, (44).

The section, p. 700, De vermibus in lapidibus, & metallis nascentibus cap. 8, contains nothing original.

Allen, Grant. The colour sense ; its origin and development. An essay in comparative psychology. 8°. London. 1879. pp. 12, 282.

In a chapter (4) on "insects and flowers," he discusses briefly the antiquity of insects and their relation to the earliest entomophilous flowers. See especially pp. 38, 42-46, 66-69, 78-80. See also Wallace, A. R.

André, Édouard. See **Brongniart, C. J. E.**

Assmann, August. Palaeontologie. Beiträge zur insekten-fauna der vorwelt. —Einleitung. 1. Beitrag. Die fossilen insekten des tertiären (miocenen) thonlagers von Schossnitz bei Kanth in Schlesien. II. Beitrag. Fossile insekten aus der tertiären (oligocenen) braunkohle von Nannburg am Bober. Mit einer tafel abbildungen. 8°. Breslau. 1869. pp. 1-62, taf. 1. (Zeitschr. f. entom. des vereins f. schles. insektenk., (2) 1.)

The introduction of thirty pages, gives a general review of fossil insects. I have seen only the separate edition. See same title in Section VI, and Section VI f.

Barrois, Charles. See **Scudder, S. H.**

Blanford, W. T. See **Medlicott, H. B., and Blanford, W. T.**

Brauer, Friedrich. See **Fritsch, A.**

Brodie, Peter Bellinger. Contributions to the geology of Glourestershire, intended chiefly for the use of students. (Geologist, [1]: 41-48, 81-88, 227-233, 289-291, 369-377.) 8°. London. 1858.

Insects are referred to on pp. 47, 231, and 375, mostly from the lias.

—— The distribution and correlation of fossil insects, and the supposed occurrence of Lepidoptera and Arachnidæ in British and foreign strata, chiefly in the secondary rocks, . . . being a paper read at the annual meeting of the Warwickshire natural history and archæological society, held at the museum, Warwick, April 15, 1873. 8°. Warwick. [n. d.] pp. 19. (Ann. rep. Warw. nat. hist. arch. soc. 37 : 12-28.) 8°. Warwick. 1873.

A very general account of fossil insects, followed (pp. 12-19) by a Tabular view of British and foreign fossil insects chiefly in the secondary rocks, omitting the foreign tertiaries. Only separate paper seen.

—— The distribution and correlation of fossil insects, and the supposed occurrence of Lepidoptera and Arachnidæ in British and foreign strata, chiefly in the secondary rocks, . . . being a paper read at the annual meeting of the Warwickshire naturalists' and archæologists' field club, held at the museum, Warwick, February 24, 1874. 8°. Warwick. [n. d.] pp. 23. (Ann. rep. Warw. nat. arch. field club. 1874. pp. 16-38.)

An enlargement of the last, principally in the tabular view. Like the preceding, it is much disfigured by typographical errors.

Brodie, P. B.—Continued.

On fossil Arachnidæ, including spiders and scorpions. 16°. Warwick. 1–2, pp.8.

A general account of what is known of fossil spiders, read before the Warwickshire naturalists' field club in March, 1882.

REPRINT. 8°. Warwick. 1883, pp. 13. Contains slight additions.

—— On the character, variety, and distribution of the fossil insects in the palæozoic (primary), mesozoic (secondary), and cainozoic (tertiary) periods: with an account of the more recent discoveries in this branch of palæontology up to the present day. 16°. Warwick. [1–90.] pp.22.

Read to the Warwickshire naturalists and archæologists' field club in March, 1889. It is of a general nature, and is remarkable as having been read more than half a century later than his first paper treating of fossil insects.

—— See also **Murchison, R. I.**

Bromell, Magnus. Lithographiæ svecanæ continuatio. Specimen II.—Sectio II. De animalibus fossilibus, illorumque variis partibus petrificatis.—Caput primum. De lapidibus insectiferis & tubulis vermicularibus.—Articulus primus. De lapidibus insectiferis scanicis & gothicis. (Acta liter. Sveciæ, 2, 493–497, 524–533, figs.) 4°. Upsaliæ et Stockholmiæ. 1729.

A general nature of the discovery of insects, "Scarabæi" and "Papiliones," in rocks of Karabylonga, Gäratad, Kuista, Olstorp, and Aklinge in Westrogothia.

Brongniart, Charles J. E. Recherches pour servir à l'histoire des insectes fossiles. Les hyménoptères fossiles. Fasc. 1. 8°. Paris. 1891. pp. 22.

Forms an Annexe on Species des hyménoptères d'Europe par M. Éd. André. The single livraison published contains introductory remarks on the rôle of insects in the world, and the mode of their preservation in a fossil state, followed by a list of the palæozoic, triassic, and liassic insects. No special reference is made to Hymenoptera, excepting in a brief note at the end.

—— See also **Girard, M.**

Bronn, Heinrich Georg. Lethæa geognostica, oder Abbildungen und beschreibungen der für die gebirgs-formationen bezeichnendsten versteinerungen mit lithographirten 47 quart-, 1 folio-tafel und 2 tabellen. 2 v. 8°. Stuttgart. 1835–38; vol. 1, pp. 6,768; vol.2,pp.[4],769–1346 [4].

Contains references to fossil insects on pp. 210, 481, 791, 800–814, 1159 1161. A second edition, which I have not seen, was published in 1836. The third was by Bronn, H. G., and Roemer, F., q. v.

Bronn, H. G.—Continued.

Index palæontologicus oder Uebersicht der bis jetzt bekannten fossilen organismen, unter mitwirkung der . . H. P. Göppert und Herm. v. Mayer, bearbeitet von Dr. H. G. Bronn. Erste abtheilung. A. Nomenclator palæontologicus, in alphabetischer ordnung. 8°. Stuttgart. 1848. pp. 6, 54, 1382. Zweite abtheilung. B. Enumerator palæontologicus; systematische zusammenstellung und geologische entwickelungsgesetze der organischen reiche. 8°. Stuttgart. 1849. t. p. pp. 980.

B (pp.585 632) refers to the geological distribution of insects.

This also appeared as Bd. 3 of the author's Handbuch der geschichte der natur. 8°. Stuttgart. 1841–'49.

—— Essai d'une réponse à la question de prix proposée en 1850 pour le concours de 1853, et puis remise pour celui de 1856, savoir: Étudier les lois de la distribution des corps organisés fossiles dans les différents terrains sédimentaires suivant l'ordre de leur superposition. Discuter la question de leur apparition ou de leur disparition successive ou simultanée. Rechercher la nature des rapports qui existent entre l'état actuel du règne organique et ses états antérieurs. (Suppl. comptes rend. acad. sc., 2: 377–918.) 4°. Paris. 1858.

See the next.

—— Untersuchungen über die entwickelungs-gesetze der organischen welt während der bildungszeit unserer erdoberfläche. Eine von der französischen akademie im jahre 1857 gekrönte preissschrift, mit ihrer erlaubniss deutsch herausgegeben. 8°. Stuttgart. 1858. pp. 10, 502.

German text of the preceding. The insects will be found treated in the original edition of this justly celebrated essay on pp. 438–53, 634–38, 810–12, 865–69.

—— See also **Gerstaecker, C. E. A.**

Bronn, Heinrich Georg, und Roemer, Ferdinand. H. G. Bronn's Lethæa geognostica oder Abbildung und beschreibung der für die gebirgs-formationen bezeichnendsten versteinerungen. Dritte stark vermehrte auflage. Mit einem atlas von 124 tafeln. 3 v. 8°. Stuttgart, 1851–1856. Atlas fol.

Bd. 1, pp. 12, 204, 6, 768,—1851–1856; Bd. 2, pp. 8, 124, 4, 570,4 412,—1851–'52; Bd. 3, pp. 8, 1130,—1853–'56. Atlas, title, schema, pl. 1–63,—1850–'56. Insects are discussed or tabulated, and typical

Bronn, H. G.—Continued.

species described, none of them new, in I, 1, 42–51; ii, 13, 75, 679–684; II, IV: 32, 429–430, III, VI: 52, 86,622 650, and figured pl. 92, 41[1a] 42[1].

Brückmann, Franciscus Ernestus. De fabulosissimae originis lapide, arachneolitho dicto, epistola ad virum clarissimum...Albertum Ritterum. 4°. Wolffenbüttelae. 1722. pp. 16, pl. 1.

Not seen; referred to by Kundmann.

—— Thesaurus subterraneus, ducatus Brunsvigii, id est: Braunschweig mit seinen unterirdischen schätzen und seltenheiten der natur. 4°. Braunschweig. 1728. pp. (4), 156, pl. 25.

On pp. 100–101 under the heading: Von denen tubulis vermicularibus des closters St. Marienthal, certain tubes composed of globular pellets are referred to water-insects and figured on pl. 19.

Brullé, Auguste. Sur le gisement des insectes fossiles, et sur les secours que l'étude de ces animaux peut fournir à la géologie. Thèse pour le doctorat ès-sciences. 4°. Paris. 1839. pp. [4], 30.

A studied review of the knowledge at that time, from which the conclusion is drawn that nearly all fossil insects are generically, and part of them specifically, identical with living types, and that in these particulars they agree with other fossil animals.

Buckland, William. Geology and mineralogy considered with reference to natural theology. [Bridgewater treatise.] 2 vols. 8°. London. 1837. Vol. 1, pp. 16, 619;—vol. 2, pp. 7, 111, pl. 1–69 (= 88 pl.).

The same. 2 vols. 8°. Philadelphia. 1837. Vol. 1, pp. 443;—vol. 2, pp. 131, pl. as in above.

See the next.

—— Geologie und mineralogie in beziehung zur natürlichen theologie... Aus dem englischen, nach der zweiten ausgabe des originals, übersetzt und mit anmerkungen und zusätzen versehen von Dr. Louis Agassiz. 2 vols. 8°, Bern, Chur und Leipzig. 1838.

Vol. 1 [text]: 1 p., pp. 26, 508; vol. 2 [plates]; 1 p., pp. 4, pl. 1–69 [88 pl.] and from 1–10 pp. of explanation of each. Pl. 46[1] and 46[1] are devoted to fossil insects, mostly arachnids, copied from Corda. The others are insects from Coalbrook-Dale, Stonesfield, and Aix. The brief text concerning them is found in the London edition at 1: 406–413, and II: 74–79; in the American edition at 1: 306–311, and ii: 74–79; in the German at pp. 453–463. The additions to the insects by Agassiz consist of a couple of unimportant notes. A new edi-

Buckland, W.—Continued.

tion by Frank Buckland (London, 1858) I have not seen, and the London edition examined is the second, apparently agreeing in every respect with the first, published in 1836.

—— Notices relative to palaeontology...from his anniversary address to the geological society of London. (Ann. mag. nat. hist., 9: 156–67.) 8°. London. 1842.

[Address in full.] (Proc. geol. soc. Lond., 3: 469–540.) 8°. London. 1841.

Contains fossil arachnidans (with opinions of Gray (J. E.) quoted), pp. 162–163 (pp. 504–505). Fossil insects, p. 163 (p. 505). Notices the arachnids described by Corda as well as specimens from Solenhofen and Aix; and the discovery of various insects the previous year from the wealden, Stonesfield slate, and Staffordshire coal, together with Hymenoptera from coal near Glasgow.

Buckman, James. See **Murchison, R. I.**

Buckton, George Bowdler. Monograph of the British Aphides. 4 vol. 8°. London. 1875–1883.

See same title in Section VII2.

Burmeister, Hermann. Geschichte der schöpfung. Eine darstellung des entwicklungsganges der erde und ihre bewohner. 8°. Leipzig. 1843. pp. 6, 487.

Refers to insects on pp. 430, 445–46. The five subsequent editions not examined.

Cope, Edward Drinker. On the evolution of the vertebrata, progressive and retrogressive. (Amer. nat., 19: 140–148, 234–247, 341–353.) 8°. Philadelphia. 1885.

A note on p 142 discusses the ancestral line of the arthropods.

Cronstedt, Axel Fredric. An essay towards a system of mineralogy. Translated from the original Swedish with notes by Gustav von Engestrom. To which is added a Treatise on the pocket laboratory, containing an easy method, used by the author, for trying mineral bodies, written by the translator. The whole revised and corrected, with some additional notes by Samuel Mendes Da Costa. 16°. London. 1780. 1 p., pp. 36, 329).

Refers, p. 264, to fossil insects found in the alum slate at Andrarum in Skåne; I also find p. 257 of the "old ed." referred to, but have been able to examine neither it nor the original Swedish.

Czech, Carl. Ueber die entwickelung des insectentypus in den geologischen perioden. (Programme realschule Düsseldorf, 1858, 1–14.) 16°. Düsseldorf. 1859.

Mainly devoted to showing that the insects of the carboniferous period were not less completely developed than the existing forms.

Da Costa, Samuel Mendes. See Cronstedt, A. F.

Dallas, William S. See Müller, F.

Dana, James Dwight. Manual of geology; treating of the principles of the science with special reference to American geological history. Illustrated by over eleven hundred and fifty figures in the text, twelve plates, and a chart of the world. Third edition. 8°. New York. 1880. pp. 14, 912, (4), pl. 12, map.

Insects mentioned on pp. 273, 274, 334–336, 342, 343, 350, 351, 388, 411, 416; many figures of American, especially paleozoic, species given. The first edition (1862) gave much less space to insects; the second (1874) does not differ from the third, as regards the insects.

Davila, Pedro Franco. Catalogue systématique et raisonné des curiosités de la nature et de l'art qui composent le cabinet de M. Davila. Tome 3. 8°. Paris. 1767. pp. 6, 290, pl. 8 [in 1st part].

On pp. 223–24. Pétrifications animales de la septième classe. Entomolites.

Dawson, Sir John William. The chain of life in geological time ; a sketch of the origin and succession of animals and plants. 16°. London. [1880.] pp. 16, 272, illustr.

In the chapter on the first air breathers a considerable number of insects are mentioned and figured, pp. 139–151, figs. 123, 126–132, including for the first time Prodryas, a fossil butterfly from Colorado.

Defrance, Jacques Louis Martin. Insectes (foss.). (Dict. sc. nat., 23: 524–526.) 8°. Paris. 1822.

A review of the older authors, questioning the validity of many of the fossils preserved in the rocks, although accepting those entombed in amber.

Demole, Isaac. See Heer, O.

Eaton, Alfred Edwin. A monograph on the ephemeridæ. (Trans. entom. soc. Lond., 1871, 1–164, pl. 1–6.) 8°. London. 1871.

Contains a chapter on fossil ephemeridæ, pp. 38–40, and a figure, pl. 1, fig. 16, of a single unnamed species from Solenhofen.

Esper, Eugen Johann Christoph. Ad avdiendam orationem pro capessendo munere philosophiæ professoris pvblici extraordinarii a rectore academiæ . . . Christiano Friderico Carolo Alexandro . . . gratiosissime sibi collato d. martii, 1783, veitandam omni qva decet observantia invitat simvlqve de animalibus oviparis et sanie frigida præditis in cataclysmo qvem svbiit orbis terrarum plerisqve salvis disserit Evgen. Joann. Christoph. Esper. 4°. Erlangæ. 1783. pp. 20.

Refers in a general way to fossil insects, pp. 18–19.

Fischer, Leopold Heinrich. Orthoptera europaea. 4°. Lipsiae. 1853. pp. 20, 454, tab. 18.

Species fossiles, pp. 55–57, contains a bibliography of fossil orthoptera and a list of the species.

Fischer von Waldheim, Gotthelf. Prodromus petromatognosiæ animalium systematicæ continens bibliographiam animalium fossilium. 4°. Mosquæ. 1826. (Nouv. mém. soc. imp. nat. Mosc., 1 : 301–374 ; 2: 95–277, 447–458.) 4°. Moscou. 1829–32.

Notices a few articles on fossil insects, tom. 2, pp. 319–20, 458.

——. Bibliographia palæontologica animalium systematica editio altera aucta. 8°. Mosquae. 1834. t. p., pp. 8, 414.

Contains slight additions to the preceding, with the notices on pp. 305, 372.

Fossil insects. (Pop. sc. monthly, 21 : 567–568.) 8°. New York. 1882.

A general statement derived from Goss's papers.

Francius, Johannis. Prodromus arachnolithographiæ. (Misc. cur. acad. nat. cur., [2], 5: 462–464.) 4°. Norimbergæ. 1687.

Refers only to the use to which "lapides aranearum" are put in medicine.

Fric, Anton. See Fritsch, A.

Fritsch, Anton. Fossile arthropoden aus der steinkohlen und kreideformation Böhmens. 4°. Wien. 1882, pp. 7, pl. 2. (Mojs. u. Neum., Beitr. paläont. österr.-ung., 2: 1–7, taf. 1–2.) 4°. Wien. 1882.

Describes and figures a carboniferous ephemerid, Palingenia feistmanteli, and three beetles, wings of a Tinea, eggs of a saw fly, and cases of a phryganid from the cretaceous beds of Bohemia; a résumé of the very few known cretaceous

Fritsch, A.—Continued.

insects is added from Goss. Brauer and Fritsch both compared the may-fly, p. 3, to the living *Palingenia longicauda.*

Gadd, Pehr Adrian. Rön och undersökning, i hvad mån insecter och zoophyter bidraga til stenhärdningar. (Kongl. vet. acad. nya handl. 8: 98–106.) 16°. Stockholm. 1787.

Refers pp. 163–164 to "globuli arenacei" which he apparently considers as eggs of insects.

Geinitz, F. Eugen. Uebersicht über die geologie Mecklenburgs; nebst geologischer karte der flötzformation Mecklenburgs. 4°. Güstrow. 1885. 30 pp., 1 pl.

Contains slight notices of fossil insects.

Geinitz, Hanns Bruno. Grundriss der versteinerungskunde. 8°. Dresden und Leipzig. 1845 [also dated 1846]. pp. 10, 815, pl. 8, tabelle 1.

B. Arthrozoa, pp. 179–193, pl. 8; gives a brief general systematic account of fossil insects, with descriptions of a few forms and figures of Æschna longilolata and Œdipoda melanostica. The second edition, 8°, Leipzig, 1856, not seen; according to Hagen the insects are upon pp. 179–90.

Gentry, T. G. Curious anomaly in history of certain larvæ of Acronycta oblinita Guenée, and hints on phylogeny of Lepidoptera. (Proc. acad. nat. sci. Phila., 1875: 24–54.) 8°. Philadelphia. 1875.

Some passing references to fossil insects on pp. 34–35.

Gerstaecker, Carl Eduard Adolph. Die klassen und ordnungen der arthropoden wissenschaftlich dargestellt in wort und bild. 5ter band, erste abtheilung. Crustacea (erste hälfte) mit 50 lithographirten tafeln. 8°. Leipzig und Heidelberg. 1866–79. *Also entitled:* Die klassen und ordnungen des thierreichs wissenschaftlich dargestellt in wort und bild. Von Dr. H. G. Bronn. Fortgesetzt von A. Gerstaecker. 5ter band: Gliederfüssler (arthropoda).

Contains in the introduction to the arthropoda in general: viii. Zeitliche verbreitung, divided into: 1. Allgemeiner charakter der fossilen arthropoden, pp. 287–292. 2. Aufeinanderfolge der formen in den verschiedenen erdschichten, pp. 293–295. Published in 1866? Under the first section the author notices the extremely small number of known fossil forms as compared with living types, and their almost complete typical agreement with existing forms; insisting that even the oldest not only fall into the orders, but even into the families of insects now extant.

Gerstaecker, C. E. A. See also **Packard, A. S.**

Gervais, Paul. See **Walckanaer, C. A. et Gervais, P.**

Giebel, Christoph Gottfried. Paläozoologie; entwurf einer systematischen darstellung der fauna der vorwelt. 8°. Merseburg. 1846. pp. 8, 360.

The insects, mentioned only by generic names, are systematically treated under each period; the period of water life on pp. 58–59, the transition period on pp. 144–148, and the period of land and air life on pp. 268–288.

—— Gæa excursoria germanica; Deutschlands geologie, geonosie und paläontologie als unentbehrlicher leitfaden auf excursionen und beim selbststudium. 16°. Leipzig. 1848. pp. 8, 510, (24), taf. 24.

Brief mention of insects on pp. 100, 296, 323, 442. Blattina didyma is figured on pl. 5, fig. 26.

—— Bericht über die leistungen im gebiete der paläontologie mit besonderer berücksichtigung der geognosie während der jahre 1848 und 1849. 8°. Berlin. 1851. pp. (4), 821.

5. Insecten, pp. 92–95, is mostly taken up with a notice of the first volume of Heer's Oeningen insects.

—— Deutschlands petrefacten; ein systematisches verzeichniss aller in Deutschland und den angrenzenden ländern vorkommenden petrefacten, nebst angabe der synonymen und fundorte. 8°. Leipzig. 1852. pp. 13, 706.

Arachniden, pp. 634–636; Insecta, pp. 636–656. A simple list.

—— Allgemeine palaeontologie; entwurf einer systematischen darstellung der fauna und flora der vorwelt; zum gebrauche bei vorlesungen und zum selbstunterrichte. 8°. Leipzig. 1852. pp. 8, 413.

Insects treated on pp. 117–118, 204–208, 276–286, under the same general divisions as in the author's Paläozoologie. The genera are enumerated.

—— Beiträge zur palaeontologie. 8°. Berlin. 1853. pp. 4, 192, pl. 3. (Jahresb. naturw. ver. Halle, 5: 287–478.) 8°. Berlin. 1853.

Contains: i. Die palaeontologie Deutschlands auf ihrem gegenwärtigen standpuncte, pp. 1–71 [287–357]. A tabular view of the genera found in Germany with the number of species of each includes, pp. 63–66 [349–352], the insects, 169 genera, and 377 or more species.—v. Bericht über den fortschritt der paläontologie während der jahre 1850–52, pp. 108–192. An analysis of the literature on fossil insects will be found on pp. 124–126 [410–412].

Giebel, C. G.—Continued.

—— Die insecten und spinnen der vorwelt mit steter berücksichtigung der lebenden insecten und spinnen : monographisch dargestellt. (Fauna der vorwelt mit steter berücksichtigung der lebenden thiere. 2er band : Gliederthiere ; erste abtheilung : Insecten und spinnen.) 8°. Leipzig. 1856. pp. 18, 511.

A systematic treatment of all the fossil insects then known with descriptions of nearly all ; many are described and named for the first time from published plates. Notice especially the treatment of the illustrations of Brodie's fossil insects of England. Some new amber Insects also appear. See also Schlechtendal, D. von, in Section VI.

—— Geologische übersicht der vorweltlichen insecten. (Zeitschr. gesammt. naturw., 8: 174-188.) 8°. Berlin. 1856.

A general review of authorities, with lists of the species mentioned in their works.

G[irard], M[aurice]. Les articulés fossiles. (La nature, 5: 301-302.) 4°. Paris. 1877.

Brief notice of recent papers by Brongniart, including that on Protomyia oustaleti.

Goss, Herbert. Introductory papers on fossil entomology. No. 1. On the importance of an acquaintance with the subject ; its bearing on the question of the evolution of insects, and the evidence it affords of the antiquity of their family types. (Entom. monthl. mag., 15: 1-5.) 8°. London. 1878.

The same. No. 2. The comparative age of the existing orders of insects, and the sequence in which they appeared on the geological horizon. (Entom. monthl. mag., 15: 52-56.) 8°. London. 1878.

The series of twelve papers of which the above form the first two covers much the same ground as the earlier series of three given in Section VI ; but the formations are followed in an ascending order, and the progress of insect life at each epoch is compared to that of other contemporary animals and plants. The lists of the other series are omitted, and the references to insects are mostly by genera.

The other papers of the series are entered under Section II, Section IV, and Section VI.

—— The geological antiquity of insects. Twelve papers on fossil entomology, reprinted, with some alterations and additions, from vols. xv and xvi of the Entomologist's monthly magazine. 8°. London. 1880. pp. (2), 50.

The preceding series, collected into a pamphlet.

Goss, H. See also **Bargagli, P. ;** **Scudder, S. H. ;** and **Fossil insects.**

Gray, John Edward. See **Buckland, W.**

Guérin-Méneville, Félix Édouard. Insectes fossiles. (Dict. classique hist. nat., 8: 579-581.) 16°. Paris. 1825.

A review of past writers, containing nothing new excepting an attempt to indicate the genera of amber insects figured by Sendel.

Haeckel, Ernst Heinrich Philipp August. Allgemeine entwickelungsgeschichte der organismen. Kritische grundzüge der mechanischen wissenschaft von den entstehenden formen der organismen, begründet durch die descendenz-theorie. (Generelle morphologie der organismen. Allgemeine grundzüge der organischen formen-wissenschaft, mechanisch begründet durch die von Charles Darwin reformirte descendenz-theorie. Zweiter band.) 8°. Berlin. 1866. pp. 160, 462, pl. 8.

In the introduction insects are treated on pp. 94-102, and the views entertained of the primeval forms of the different groups supported in part by paleontological evidence.

—— Natürliche schöpfungsgeschichte. Gemeinverständliche wissenschaftliche vorträge über die entwickelungslehre im allgemeinen und diejenige von Darwin, Goethe und Lamarck im besonderen. Vierte verbesserte auflage. 8°. Berlin. 1873. pp. 46, 688, pl. (1), 15.

TRANSLATION: The history of creation : or the development of the earth and its inhabitants by the action of natural causes. A popular exposition of the doctrine of evolution in general, and of that of Darwin, Goethe, and Lamarck in particular. The translation revised by E. Ray Lankester. 2 vols., 12°. New York. 1876.—vol. 1, pp. 20, 374, pl. (1), 1-3 ; vol. 2, pp. 8, 408, pl. 1-15.

Insects are treated on pp. 490-501 (transl., 2: 178-191) and their pedigrees considered, partly from geological considerations.

Hagen, Hermann August. Die fossilen libellen Europas. (Stett. entom. zeit., 9: 6-13.) 16°. Stettin. 1848.

A revision and brief description of the fifteen species then known.

—— Ueber die lebensweise der termiten und ihre verbreitung. (Königsb.

Hagen, H. A.—Continued.

naturw. unterh., 2, iii: 53–75.) 8°. Königsberg. 1852.

Page 71 treats of the fossil species in amber, and from the tertiary beds of Oeningen and Radoboj, as proving a warmer climate in ancient Europe; of the sixty known species of white ants one-third were fossil.

—— Monographie der termiten. (Linn. entom., 10: 1–141, 270–325,—1855; 12: 1–342, pl. 1–3,—1858; 14: 73–128,—1860.) 8°. Stettin. 1855–60.

Includes a treatment of the (14) fossil species with the others. Besides this, under the head Literatur (palaeontologie), 10: 302–310; 12: 291–298, an analysis is given of works in which the fossil species have been previously treated. See also ⊙ in Section VIIIc.

—— Catalogue of the specimens of neuropterous insects in the collection of the British museum. Part I. Termitina. 12°. London. 1858. pp. 34.

Contains the fifteen fossil species described in the Monographie der termiten, from which indeed the whole was compiled [by Adam White] without the knowledge of the reputed author. None of the fossil species are recorded as in the collections of the British museum.

—— Synopsis pseudoscorpionidum systematica. (Proc. Bost. soc. nat. hist., 13: 263–272.) 8°. Boston. 1870.

A synonymic list of the known species of which fifty are recorded, ten of them (one, however, doubly recorded) fossil, all but one being from amber.

—— See also **Packard, A. S.**; and **de Selys-Longchamps, E., et Hagen, H. A.**

Heer, Oswald. Zur geschichte der insekten. Vortrag. 8°. n. p., n. d. pp. 20. (Verhandl. schweiz. gesellsch. gesammt. naturw., 34: 78–97.) 8°. Frauenfeld. 1849. (Neues jahrb. f. mineral., 1850, 17–33.) 8°. Stuttgart. 1850.

A popular address, presenting a sketch of the sequence of insect life and the development of special groups, with general considerations based on a broad survey of the subject; by far the best account of the knowledge of that time. An abstract is given in Haidinger's Berichte, 6: 135–136. 8°. Wien. 1849.

TRANSLATION. On the history of insects. (Quart. journ. geol. soc. Lond., 6, ii: 68–76.) 8°. London. 1850.

Translation by T. R. Jones.

Heer, O.—Continued.

—— Ueber die fossilen kakerlaken. (Vierteljahrsschr. nat. gesellsch. Zürich 9, IV: 273–302, pl.) 8°. Zürich. 1864.

The first attempt to classify the cockroaches of the carboniferous period, followed by a catalogue of the fifty-four known fossil species from all formations, and descriptions and figures of ten new species.

—— Die urwelt der Schweiz. Mit sieben landschaftlichen bildern, elf tafeln, einer geologischen übersichtskarte der Schweiz und zahlreichen in den text eingedruckten abbildungen. 8°. Zürich. 1865. pp. 20, 622, pl. 7, (4), map, 368 figs. in text.

Contains a general account of the lias insects, pp. 81–96, pl. 7–8, of those of Oeningen, pp. 355–397, figs. 215–323; and of the pleistocene of Utznach, etc., pp. 500–503, figs. 352–359. Many forms are described and figured for the first time.

TRANSLATION. Le monde primitif de la Suisse. Traduit de l'allemand par Isaac Demole. 8°. Genève et Bâle. 1872. pp. 16, 801, pl. 11, carte, 368 figs. in text.

The insects occupy p. 22, fig. 16 e (carboniferous); pp. 99–117, pl. 7–8 (lias); pp. 436–486, figs. 215–323 (Oeningen); and pp. 613–616, figs. 352–359 (Utznach, etc.). Some few additions are made by the author.

TRANSLATION. The primitive world of Switzerland, with 560 illustrations. By Professor Heer. Edited by James Heywood. 2 vols. 8°. London. 1876. Vol. 1, pp. 16, 393, map, pl. 6; —vol. 2, pp. 8, 324, pl. xi and 4 scattered plates.

The insects occupy i: p. 20, fig. 16 e (carboniferous); pp. 81–95, pl. 7–8 (lias); ii: pp. 9–56, figs. 211–323 (Oeningen); and pp. 167–170, figs. 352–358 (Utznach, etc.)

—— Die urwelt der Schweiz. Zweite umgearbeitete und vermehrte auflage. 8°. Zürich. 1879. pp. 19, 713, taf. 8, (4), map, 417 figs. in text.

The insects are here somewhat enlarged over the previous editions, occupying pp. 24–25, fig. 34 (carboniferous); pp. 91–105, pl. 7–8 (lias); pp. 380–422, figs. 256–365 (Oeningen); and pp. 530–533, figs. 395–402 (Utznach, etc.).

—— Flora fossilis arctica. Die fossile flora der polarländer. 6 v. 4°. Zürich. 1868–80. Bd. 1, 1868, pp. 7, 192, map, pl. 50;—bd. (2), 1869–71 (no t. p.), pp. 7; (i.) pp. 415–488, pl. 39–56; (ii.) pp. 41, pl. 10; (iii.) pp. 98, pl. 16; (iv.) pp. 51, pl. 15;—bd. 3, 1875, t. p., pp. 6; i. pp. 11, pl. 6; ii. pp. 138, (2), pl. 38; iii. pp. 29, pl.

Heer, O.—Continued.

5; iv. pp. 24;—bd. 4, 1877; i. pp. 7, 141, pl. 32; ii. t. p., pp. 122, pl. 31; iii. pp. 15, pl. 2.—bd. 5, 187■ i. pp. 4, 38, front., pl. 9; ii. t. p., pp. 58, pl. 15; iii. t. p., pp. 61, pl. 15; (iv.) pp. 11, pl. 4; (v.) pp. 6, pl. 1;—bd. 6. i, 1880, pp. (4), t. p., 34, 17, 38, pl. 9, 6, 3.

The contents will be found under the special papers.

——— Ueber die fossilen insekten Grönlands. (Flora foss. grönl., ii: 143-148, pl. 109 pars.) 4°. Zürich. 1883.

Mentions or describes and figures three Coleoptera from the cretaceous of Kome and Ivnangut, eight Coleoptera, two Orthoptera, one Neuropteron, and two Hemiptera from the various tertiary deposits, but mostly from Atanekerdluk.

Extracted from Grönlands geol. undersögelse, which I have not been able to consult.

Heilprin, Angelo. The geographical and geological distribution of animals. (Internat. scient. series, t.VII.) 8°. New York. 1887. pp. 12, 435. map.

Distribution of insects on pp. 270-285, Arachnida and Myriapoda, 285-286.

Heywood, James. See **Heer, O.**

Hislop, Stephen, and **Hunter, R.** On the geology and fossils of the neighborhood of Nágpur, Central India. (Quart. journ. geol. soc. Lond., 11: 345-383, pl. 10.) 8°. London. 1854.

Contains, p. 360, a reference to insects.

Holl, Friedrich. Handbuch der petrefactenkunde; mit einer einleitung über die vorwelt der organischen wesen auf der erde, von Dr. Ludwig Choulant. 1ᵉˢ bändchen. 16°. Dresden. 1820. pp. 8, 489. (Allg. taschenbibl. der naturwiss., 9ᵗᵉ theil.)

A brief account of fossil insects under the heading Entomolithen, pp. 138-141, with description of two species of Formica from amber.

——— Handbuch der petrefactenkunde; eine beschreibung aller bis jetzt bekannten versteinerungen aus dem thier- und pflanzenreiche zur leichten erkennung und auffindung der fossilien; mit einer einleitung über die vorwelt der organischen wesen auf der erde, von Dr. Ludwig Choulant. Neue ausgabe. 16°. Quedlinburg und Leipzig. 1843. pp. 8, 489. Published in four parts with continuous pagination, the t. p. of pt. 2-4 not included.

Appears to differ from the preceding only in title.

Horn, George Henry. See **Le Conte, J. L.,** and **Horn, G. H.**

Hunter, R. See **Hislop, S.,** and **Hunter, R.**

Jones, Thomas Rupert. See **Heer, O.;** and **Mantell, G. A.**

Jordan, Johann Ludwig. Mineralogische berg- und hüttenmännische reisebemerkungen, vorzüglich in Hessen, Thüringen, am Rheine und in sayn-altenkirchnerischen gebiete. 8°. Göttingen. 1803.

Not seen; said to contain, on p. 123, some reference to fossil insects.

Keferstein, Christian. Die naturgeschichte des erdkörpers in ihren ersten grundzügen dargestellt. 2 v. 8°. Leipzig. 1834. 1ᵉʳ theil, pp. 11, 394; 2ᵉʳ theil, pp. 4, 896.

Abtheilung 2 (paläontologie), sechster abschnitt: Die fossilen insekten, pp. 325-347; under 7ᵉʳ abschnitt the myriapods and arachnids, F and G, pp. 370-371, 375-376, 378. The species are enumerated in the two last-mentioned groups, but only the genera in the hexapods; the names are very frequently misspelled. See also Vollmar.

Kolbe, H. J. Die vorläufer (prototypen) der höheren insectenordnungen im paläozoischen zeitalter. (Berl. ent. zeitschr., 28: 167-170.) 8°. Berlin. 1884.

Has a few words only on paleontological evidence.

——— Einführung in die kenntnis der insekten. 8°. Berlin. 1889-1890.

The prospectus of this work, now in the course of publication, promises a section vii. entitled Die ausgestorbenen insekten (Paläontologie).

Lankester, Edwin Ray. See **Haeckel, E. H. P. A.**

LeConte, John Lawrence. Address before the American association for the advancement of science at Detroit, Michigan, August 13, 1875. 8°. Salem. 1875. t. p., pp. 18. (Proc. Amer. assoc. adv. sc., 24: 1-18.) 8°. Salem. 1876.

The distribution of certain North American beetles directly indicates a survival from cretaceous or even earlier times, pp. 4-7.

——— and **Horn, G. H.** Classification of the Coleoptera of North America. 8°. Washington. 1883. 38, 567 pp. (Smithson. misc. coll., 507.)

On p. 30 of the introduction is a succinct account of the appearance of fossil insects.

Lemoine, *Dr.* [Insectes fossiles des environs de Reims.] (Bull. soc. ent. Fr., 1887: 17.) 8°. Paris. 1887.

Exhibition of drawings; no details.

Lesley, J. Peter. A dictionary of the fossils of Pennsylvania and neighboring States named in the reports and catalogues of the survey. (Vol. 1, A–M.) 8°. Harrisburg. 1889. pp. 14, 437, 31. Figs. (Geol. surv. Penn. rep. P³.)

The known palæozoic insects are catalogued with a few of later date, their horizon indicated, and copies of illustrations given in many cases. Four hundred pages of the second volume have been printed but not yet published. The pagination is continued throughout the two volumes.

Lesser, Friedrich Christian. Lithotheologie, das ist: Natürliche historie und geistliche betrachtung derer steine. 16°. Hamburg. 1735. pp. 42, 300, (52), pl. 10.

In the seventh chapter, fourth division, fifth book: Von versteinerten thieren auf erden, so kein blut haben, pp. 353-361, he reviews what is known of fossil insects in his day.

In the Hamburg edition of 1751 (pp. 48, 1488, pl. 10) the same appears on pp. 633-639.

Lhwyd, E. See Luidius, E.

Linné, Carl von. Oeländska och gothländska resa på riksens högloflige ständers befallning förrättad åhr 1741; med anmärkningar uti oeconomien, naturalhistorien, antiquiteter, &c. med åtstilliga figurer. 16°. Stockholm och Upsala. 1745. pp. (14), 344, 30, 2 maps, pl., figs.

Contains a mere mention, p.59, of finding some small insects in a fossil state near Glimminge in Oeland.

TRANSLATION: Reisen durch Oeland und Gothland welche auf befehl der hochlöblichen reichsstände des königreichs Schweden im Jahr 1741 angestellt worden. 16°. Halle. 1764. pp. (32), 364, (24), 2 maps, 2 pl.

The same on p. 68.

Lippi. [Lettre à M. Dodart.] (Hist. acad. sc., 1705 : 36-37.) 4°. Paris. 1706.

Account of the discovery of supposed bee-cells (probably corals) in the rocks of the Montagnes de Sout, Upper Egypt.

.·. It is somewhere stated that Lippi has mentioned the fossil insects of Oeningen.

Lochnerus, Johannes Henricus et Michael Fridericus. Rariora mvsei besleriani quae olim Basilivs et Michael Ropertvs Besleri collegervnt aeneisqve tabvlis ad vivvm incisa evvigarvnt: nunc commentariolo illustrata a Johanne Henrico

Bull. 69———2.

Lochnerus, J. H. et M. F.—Continued. Lochnero, vt virtvti toy makaritoy exstaret monvmentvm, denvo lvci pvblicae commisit & laudationem ejvs fvnebrem adjecit maestissimvs parens Michael Fridericvs Lochnervs. folio. n. p. 1716. pp. (22), 112, pl. 40, portr. 2.

Not seen; according to Kundmann, contains references to fossil insects on pp. 34, 100.

——— Michael Fridericus. See Lochnerus, J. H. et M. F.

Lubbock, Sir John. The president's address. (Trans. ent. soc. Lond., [3], 5, journ. of proc., 113-131.) 8°. London. 1867.

Records, pp. 128-129, the progress during the year in the study of fossil insects.

——— Monograph of the Collembola and Thysanura. 8°. London. 1873. pp. 10, 276, pl. 78.

His second chapter, miscalled Chapter III, is On the importance of the Collembola and Thysanura in relation to the evolution of the Insecta, pp. 40-54, and takes into consideration the facts then known of the geological history of the latter. It will be found suggestive.

——— On the origin and metamorphoses of insects. 16°. London. 1874. pp. 16, 108.

Contains a chapter [v] on the origin of insects, in which, on p. 86, is a general statement of the geological appearance of the different orders of insects.

——— Address read before the entomological society of London at the anniversary meeting on the 19th January, 1881. 8°. London. 1881. pp. 17. (Trans. ent. soc. Lond., 1880, journ. of proc., 11-55). 8°. London. 1881.

Refers, pp. 12-13 (50-51), to recent researches on fossil insects, particularly by Goss and Scudder.

Luidius, Edvardus. Edvardi Luidii apud oxonienses cimeliarchae ashmoleani Lithophylacci britannici ichnographia. Sive lapidum aliorumque fossilium britannicorum singulari figura insignium, quotquot hactenus vel ipse invenit vel ab amicis accepit: distributio classica: scrinii sui lapidarii repertorium cum locis singulorum natalibus exhibens; additis rariorum aliquot figuris aere incisis: cum epistolis ad clarissimos viros de quibusdam circa marina fossilia et stirpes minerales praesertim notandis. Editio altera: novis quorumdam speciminum iconibus aucta; subjicitur authorio praelecto de stellis marinis, etc. 8°. Oxonii. 1760. pp. (16), 156, (4), pl. 25.

Luidius, E.—Continued.

Epistola 3. Sammarium Litefurum V. C. D. Richardi Richardson, M. D. De entiocho lapide, conchitis, et lithophytis seu plantis mineralibus agri choraceonsis; de bufonibus medlis saxis inclusis, et depictis aliquot in schisto carbonaria insectis, pp. 107-114, pl. 4, fig. 197 (4 figs.). First edition, Lipsiae, 1699, not seen.

Lyell, Sir Charles. Elements of geology, or The ancient changes of the earth and its inhabitants as illustrated by geological monuments; sixth edition, greatly enlarged and illustrated with 770 woodcuts. 8°. London. 1865. pp. 16, 794.

Contains references to fossil insects and some illustrations of them on pp. 243, 250, 255, 331, 425, 491. Earlier editions not examined.

McCook, Henry Christopher. American spiders and their spinning work. A natural history of the orbweaving spiders of the United States, with special regard to their industry and habits. 3 v. Imp. 8°. Philadelphia. 1889—1893-. Vol. 1, pp. 372, figs. 345, 1889;—vol. 2, pp. 480, figs. 401, pl. 5, 1890;—vol. 3 not yet published.

Vol.2 contains Part VI: Fossil spiders; chap. xv; pp. 446-469, figs. 372-401: Ancestral spiders and their habits. It gives a general review of fossil spiders, accompanied by several illustrations, with the general conclusion that the structure and habits of spiders have not greatly differed from the earliest times from which they are known.

McLachlan, Robert. Insects. (Encycl. brit., ed. 9, vol. 13, pp. 141-154.) 4°. Edinburgh [and Boston]. 1881.

Contains a paragraph, p. 141, on the Antiquity of insects.

Mantell, Gideon Algernon. The wonders of geology. First American from third London edition, 2 vols. 16°. New Haven. 1839. Vol. 1, pp. 16, 1-428, front., pl. 4;—vol. 2, pp. 7, 429-804, (24), pl. 6-10.

"Fossil insects" (of Aix). 1: 247-250, tab. 45. "Insects of the coal formation." 2: 679-680.

——— *The same:* 4th ed. London. 1840. 6th ed. 2 v. 16°. London. 1848. Vol. 1, pp. 15, 452;—vol. 2, pp. 453-938, plates as above. 7th ed. revised and augmented by T. Rupert Jones. 2 v. 16°. London. Vol. 1, pp. 24, 1-480, (1857);—vol. 2, pp. 16, 481-1019 (1858).

——— The medals of creation, or First lessons in geology and in the study of organic remains. 2 vols. 16°. London. 1844. Vol. 1, pp. 28, 1-456, pl. 1, 3-6;—vol. 2, pp. 6, 457-1016, pl. 2.

Fossil insects and spiders, pp. 570-581, with wood-cuts 122-124.

Mantel, G. A.—Continued.

——— *The same:* 2d edition entirely rewritten. 2 v. 16°. London. 1854. Vol. 1, pp. 321-446;—vol. 2, pp. 11, 447-930, plates as before.

——— Geological excursions round the Isle of Wight and along the adjacent coast of Dorsetshire; illustrative of the most interesting geological phenomena, and organic remains. 16°. London. 1847. pp. 428, pl. 20.

Refers to the discovery of fossil insects in tertiary and wealden beds at pp. 119, 400.

Marcou, John Belknap. A review of the progress of North American invertebrate palaeontology for 1883. (Amer. nat., 18: 385-392.) 8°. Philadelphia. 1884.

Notices two or three papers on fossil insects.

The same for 1884. (Ibid., 19: 353-360.) 8°. Philadelphia. 1885.

Notices Scudder's papers on fossil insects.

——— A review of the progress of North American palaeontology for the year 1884. 8°. Washington. 1885. 2, 20 pp. (Rep. Smiths. inst., 1884, 563-582.) 8°. Washington. 1885.

Contains brief notices, among others, of about twenty papers on fossil insects.

Maurice, Charles. Les insectes fossiles spécialement d'après les travaux de Sir [sic] Samuel Scudder. 8°. Lille. 1882. pp. 1-31. (Ann. soc. géol. Nord, 9: 152-183.) 5°. Lille. 1882.

A general account of what is known of fossil insects, arranged in geological succession.

Medlicott, H. B., and Blanford, W. T. A manual of the geology of India; chiefly compiled from the observations of the geological survey; published by order of the government of India. 8°. Calcutta. 1879. 2 vol. and map. Vol. 1. Peninsular area, pp. 18, 80, 1-444;—vol. 2. Extrapeninsular area, pp. 445-817, pl. 21.

On pp. 152, 154, 314 are references to insects found in the Mesozoic and Tertiary deposits of central India.

Meyer, Christian Erich Hermann von. Insekten, fossile. (Ersch u. Grüber, Allg. encycl. wissensch. u. kunste, sect. 2, th. 18, s. 536-541.) 4°. Leipzig. 1840.

A review of the literature, with nothing new.

——— See also **Reuss, A. E.,** und **Meyer, C. E. H. von.**

Morris, John. A catalogue of British fossils, comprising all the genera and spec-

Morris, J.—Continued.

ies hitherto described ; with references to their geological distribution and to the localities in which they have been found. 8°. London. 1843. pp. 11, 222.

Insecta, p. 69.

——— *The same.* Second edition, considerably enlarged. 8°. London. 1854. pp. 8, 372.

Insecta, pp. 116-118.

The recent third edition not examined.

Mourlon, M. Géologie de la Belgique. 2 vol. 8°. Bruxelles, 1880-1881. vol. 1, (1880), pp. 4, 317 ;—vol. 2,(1881), pp. 4, 16, 392.

Refers in three brief paragraphs (1: 125, 144) to the insects reported from the coal by van Beneden and de Borre, and from the oolite by the latter ; the carboniferous species are also catalogued (2: 57) as well as larvæ of insects from the wealden of Hainaut (2: 82).

Müller, Fritz. Facts and arguments for Darwin ; with additions by the author; translated by W. S. Dallas. 16°. London. 1869? pp. (8), 144.

Argues in favor of the late acquisition of "complete" metamorphoses in insects partly from palæontological data, in a foot-note to pp. 119-121 ; it does not occur in the original, entitled Für Darwin.

Münster, Georg, *graf zu.* Fossile fische, sepien, krebse, Monotis salinaria, saurier, algaciten von Oeningen, schildkröte in lias von Altdorf, Clymenia glossopteris, folliculites ; tertiär formation in Norddeutschland. (Neuesjahrb. mineral., 1834: 42-43.) 8°. Stuttgart. 1834.

Mentions the reception of a beetle from Solenhofen and insects from Oeningen.

Murchison, *Sir* Roderick Impey. Outline of the geology of the neighborhood of Cheltenham. A new edition, augmented and revised by James Buckman and H. E. Strickland. 8°. London. 1845. pp. 110, map, sect., pl. 13.

The first edition makes no allusion to fossil insects. The additions upon that subject in this are by Brodie (q. v. in Section IV) and Buckman. The insects are catalogued on pp. 65, 81 82, and figured on pl. 4, 8, 9. Reference is also incidentally made in several places to the insect beds of the district, but without special mention of their contents.

Mylius, G. F. Memorabilium Saxoniæ subterraneæ, pars prima : i. e. Des unterirdischen Sachsens weltsamer wunder der natur ; erster theil. Worinnen die auf denen steinen an kräutern, bäumen, blüh-

Mylius, G. F.—Continued.

men, fischen, thieren und anderen dergleichen besondere abbildungen, so wohl unsers Sachsenlandes als deren so es mit diesen gemein haben, gezeiget werden, mit vielen kupfern gezieret. 4°. Leipzig. 1709. 1qe. (6), 80, (19), front., pl. (13), fig.

Remarks on a fossil "wurm, welcher einem seidenwurm nicht ungleich scheinet." p. 56.

Nicholson, Henry Alleyne. A manual of palæontology for the use of students, with a general introduction on the principles of palæontology. 8°. London. 1872. pp. 12, 601, figs. 401.

Chapter xvii, Arachnida, Myriapoda, and Insecta, occupies pp. 181-187 and figs. 124-130. This very brief notice is mostly confined to the older insects.

——— *The same.* Second edition, revised and greatly enlarged. 2 vol. 8°. Edinburgh and London. 1879. Vol. 1, pp. 12,511 ;—vol. 2, pp. 12, 531, figs.

Vol. 1 contains a chapter (20) on Arachnida, Myriapoda, and Insecta, pp. 398-409, figs. 250-258 ; slightly enlarged from the preceding. See also Nicholson, H. A., and Lydekker, R.

——— *The ancient life-history of the earth, a comprehensive outline of the principles and leading facts of palæontological science.* 8°. Edinburgh and London. 1877. pp. 19, 407.

Contains nothing original in insects ; a few are figured.

——— See also **White, C. A.,** and **Nicholson, H. A.**

——— and **Lydekker, Richard.** A manual of palæontology for the use of students, with a general introduction on the principles of palæontology. 3d edition, rewritten and greatly enlarged. Two vols. 8°. Edinburgh and London. 1889. Vol. 1, pp. 12, 1-885, figs. 1-812 ;—vol. 2, pp. 11, 886-1624, figs. 813-1419.

Chapt. 31, pp. 572-586, figs. 430-440, and chapt. 32, pp. 587-602, figs. 441-449, deal respectively with the arachnida and myriapoda, and the insects. The figures, when of fossils, are all borrowed, excepting one showing the skin structure of a Carboniferous scorpion. This is one of the best summary accounts of fossil insects which has been published. The invertebrata are all by Nicholson alone.

Oppenheim, Paul. Die ahnen unserer schmetterlinge in der sekundär-und tertiärperiode. (Berl. entom. zeitschr., 29: 331-349, [3] pl.) 8°. Berlin. 1885.

Describes Palæocossus jurassicus from the brown Jura of Siberia, and characterizes the

Oppenheim, P.—Continued.
Rhipidorhabdi of the Bavarian white Jura with two genera and six species; these he regards as the ancestors of later Lepidoptera. A list of known tertiary Lepidoptera is appended, grouped by strata; a single species is described.

d'Orbigny, Alcide Dessalines. Recherches zoologiques sur l'instant d'apparition, dans les âges du monde, des ordres d'animaux, comparé au degré de perfection de l'ensemble de leurs organes. (Ann. sc. nat., (3), 13: 228–236, 2 tables.) 8°. Paris. 1850.
Insects are treated on pp. 232–233, and on both tables.

Oustalet, Émile. Paléontologie. (Girard, Les insectes; traité élémentaire d'entomologie, 1: 170–180.) 8°. Paris, 1873.
A good general account of fossil insects from the older to the newer strata.

Owen, Sir Richard. Palæontology, or a systematic summary of extinct animals and their geological relations. 2d edition. 8°. Edinburgh. 1861. pp. 16, 463.
The insects are briefly treated, without illustrations, on pp. 51–52. Nothing new is given except the expression of a doubt by Waterhouse whether the indusial limestone of Auvergne is to be referred to the cases of Phryganidæ.
The first edition of this work was an excerpt from vol. 17 of the Encyclopædia Britannica (8th ed., 1859), where the article occupied pp. 91–176, and the insects, in nearly the same words as in the later edition, occurred on pp. 102, 103.

Packard, Alpheus Spring. Guide to the study of insects, and a treatise on those injurious and beneficial to crops: for the use of colleges, farm-schools, and agriculturalists; with eleven plates and six hundred and fifty wood-cuts. 8°. Salem. 1869 [1868–69], pp. 8, 702, pl. 11.
Has a section on geological distribution, pp. 77–81, with a plate (1) in the introduction, besides treating of the fossil species in the body of the work; especially in the Neuroptera, where pp. 582–584 are given to a discussion of Engerson, with a figure (572) and quotations from opinions expressed by Hagen and Gerstaecker in letters from the former. In the third edition (1872) an appendix is added, in which, pp. 710–711, a description and figure are given of Paola. A figure of Arthrolycosa is added in the fourth edition (1874). The preface to sixth edition (1875) contains on its second page a paragraph on the fossil insects published in America since the previous edition [furnished by S. H. Scudder].

———. Our common insects: a popular account of the insects of our fields,

Packard, A. S.—Continued.
forests, gardens, and houses; illustrated with four plates and 268 wood cuts. 16°. Salem. 1873.
Contains a chapter (xiii) entitled Hints on the ancestry of insects, in which, and especially on pp. 157–159, the geological question is briefly discussed.

Parkinson, James. Organic remains of a former world; an examination of the mineralized remains of the vegetables and animals of the antediluvian world; generally termed extraneous fossils. 2d edition. 3 vols. 4°. London. 1833. Vol. 1, pp. 12, 460, (4), front., pl. 1, 9;—vol. 2, pp. 14, 286, (26), front., pl. 19;—vol. 3, pp. 12 [10], 467, (20), front., pl. 22.
Entomolithi are treated in vol. 3, pp. 265–267, and pl. 17, figs. 2–9; larvæ of Odonata from Pappenheim being figured and some indistinguishable insects copied from Luidius.

Phillips, John. See **Woodward, H. et al.**

Pictet-Baraban, afterward Pictet de la Rive, François Jules (q. v.).

Pictet de la Rive, François Jules. Traité élémentaire de paléontologie ou Histoire naturelle des animaux fossiles considérés dans leurs rapports zoologiques et géologiques. 4 vol. 8°. Genève. 1844–46. T. 4. (1846), pp. 15, 458, pl. 20 in text.
IVe classe, Arachnides, pp. 87–89; V^e classe, Insectes, pp. 91–114; VIe classe, Myriapodes, pp. 115–116. No insects are figured.

———. Traité de paléontologie ou Histoire naturelle des animaux fossiles considérés dans leurs rapports zoologiques et géologiques. 2e édition. 4 vol. 8°; atlas, 4°. Paris. 1853–57. T. 2 (1854), t. p., pp. 727; atlas, pp. 32, pl. 56.
Insectes, pp. 301–405; Myriapodes, p. 405; Arachnides, pp. 406–410; Atlas, pl. 40–41. A few additions from Serres MS. are quoted.

Pidgeon, Edward. The fossil remains of the animal kingdom. 8°. London. 1830. pp. (6), 514, (1), pl. (49).
Forms the supplementary volumes of Griffith's Animal kingdom of Cuvier. Insects are treated in a summary manner on pp. 493–495, but nothing new is added, and none are figured.

Plinius Secundus, Caius. Naturalis historia. Liber 37, section 72.
As translated by Bostock and Riley (8°, London, 1857) that part of the passage which may re-

Plinius Sec., C.—Continued.

fer to fossil insects reads as follows: "Other stones, again, derive their names from various animals * * * scorpitis, from either the color or the shape of the scorpion * * * Myrmecitis presents the appearance of an ant crawling within, and cantharias of a scarabæus."

Quenstedt, Friedrich August. Handbuch der petrefaktenkunde. 8°. Tübingen. 1852. pp. 6, 792, pl. 62.

Arachnidæ, pp. 306-309; Insecta, pp. 309-319. The 2d edition not seen.

Quinet, Edgar. La création. 3e éd. 2 v. 12°. Paris. 1879. (Œuvres compl., xxi-xxii) vol. 1, 2 t. p., pp. 6, 359;—vol. 2, 2 t. p., pp. 382.

The first three chapters of book v are devoted to a popular and picturesque though often faulty account of the geological development of insects. The work has been translated into German by Cotta and passed through several French editions, the first in 1869 (?).

Reuss, August Emanuel. Die gegend zwischen Kommotau, Saaz, Raudnitz und Tetschen. (Löschn., Beitr. balneol., 2.) 8°. Prag. 1847.

Not seen; according to Deichmüller, this article contains some reference on p. 46 to the occurrence of fossil insects in certain Bohemian localities.

—— und **Meyer, C. E. H. von.** Die tertiären süsswassergebilde des nördlichen Böhmen's und ihre fossilen thierreste. (Palaeontogr., 2: 1-73, pl. 1-12.) 4°. Cassel. 1849-51.

Contains Geognostische skizze der tertiären süsswasserschichten des nördlichen Böhmens, pp. 1-15, by Reuss alone, in which are recorded the discovery of Coleoptera at Kutschen (p. 6), and of insects, principally Coleoptera, at Luschitz (p. 7).

Roemer, Ferdinand. Lethæa geognostica oder Beschreibung und abbildung der für die gebirgs-formationen bezeichnendsten versteinerungen, herausgegeben von einer vereinigung von paläontologen. I. theil: Lethæa palaeozoica von F. Roemer. With secondary title: Lethæa geognostica oder Beschreibung und abbildung der für die einzelner abtheilung der palaeozoischen formation bezeichnendsten versteinerungen. Atlas, mit 62 tafeln. 8°. Stuttgart. 1876. 2 t. p., vorw. 2 p., taf. 62 (mit erläuterungen.)

A few insects are figured on plates 31, 47, and 56.

—— See also **Bronn, H. G.,** und **Roemer, F.**

Scheuchzer, Johann Jacob. Meteorologia et oryctographia helvetica; oder Beschreibung der luft-geschichten, steinen, metallen, und anderen mineralien des Schweitzerlands, absonderlich auch der überbliebselen der sündfluth. Ist der dritte oder eigentlich der sechste theil der Natur-geschichten des Schweitzerlands. 4°. Zürich. 1718. t. p., ff. 7, pp. 336, pl. (19).

Under Insecta diluviana on p. 350 is a paragraph saying that while the author is acquainted with fossil insects he has seen none from Switzerland; quotes Langius.

Schlotheim, Ernst Friedrich von. Die petrefactenkunde auf ihrem jetzigen standpunkte durch die beschreibung seiner sammlung versteinerter und fossiler überreste des thier- und pflanzenreichs der vorwelt erläutert; mit 15 kupfertafeln. 8° (atlas, 4°). Gotha. 1820. pp. 64, 438.

Pp. 42-44 relate to insects, specified under six heads.

—— Nachträge zur petrefactenkunde; mit 21 kupfertafeln. 8°. Gotha. 1822. pp. 12, 100.

The same. Zweyte abtheilung; mit kupfertafeln. 8° (atlas, 4°). Gotha. 1823. pp. (4), 114.

The second part contains on pp. 60-61, taf. 22, fig. 10, what the author looks upon as the larva and nest of a Myrmeleon.

Schreber, Johann Christian Daniel. Lithographia halensis exhibens lapides circa Halam Saxonvm reperivndos systematice digestos secvndvm classes et ordines genera et species cvm synonymis selectis et descriptionibvs speciervm. Praefatvs est Ioh. Ioach Langivs. 16°. Halae. 1759. pp. 24, 80, pl. 1.

Describes Entomolithus coleopteri from Rothenburg, pp. 51-52.

Schröter, Johann Samuel. Entomolithen, versteinte insecten (Schröter, Lithol. real- u. verbal-lex., 2: 93-100.) 8°. Frankfurt. 1779.

A compilation from the writings of the older authors.

—— Insecten, entomolithen. (Schröter, Lithol. real- u. verballex., 3: 72-75.) 8°. Frankfurt. 1780.

General remarks of no present value.

—— Neue litteratur und beyträge zur kenntniss der naturgeschichte vorzüglich

Schröter, J. S.—Continued.

der conchylien und fossilien. 1er band. 16°. Leipzig. 1784. pp. (8), 550, (30), pl.3.

V. Ueber einige merkwürdige versteinerungen. A. Von versteinten insecten, 410–413, pl. 3, fig. 16.

Scudder, Samuel Hubbard. The insects of ancient America. (Amer. nat., 1: 625–631, pl. 16.) 8°. Salem. 1868.

A popular account of those then known.

—— The fossil insects of North America. Geol. mag., 5 : 172–177, 216–222.) 8°. London. 1868.

A review of the 87 species then known and of their geological relations. Abstracts will be found in the Amer. nat., 1: 557. 8°. Salem. 1867;—Can. nat., (2), 3: 293–294. 8°. Montreal. 1868;—Quart. journ. sc., 5: 400. 8°. London. 1868;—and Pop. sc. rev., 7: 316–319. 8°. London. 1868.

—— Entomological notes, V. 8°. Boston. 1876. pp. 72.

Reprints, with other matter, two papers containing notes on fossil insects, mentioned elsewhere in this bibliography.

—— Some recent publications on fossil insects. (Psyche, 3: 138.) 8°. Cambridge. 1880.

A review of Goss's papers.

—— A bibliography of fossil insects. (Harv. univ. bull., 2: 48–51, 87–88, 122–124, 157–162, 202–208, 252–257, 296–299, 333–337, 407–411.) 4°. Cambridge. 1881–1882. (Bibl. contr. libr. Harv. univ., 13.) 47 pp. 4°. Cambridge. 1882.

"More than four hundred authors are represented as having concerned themselves more or less with fossil insects."

—— [Minor notices of fossil insects.] (Psyche, 3: 277–279.) 4°. Cambridge. 1882.

Exhibition at meetings of the Cambridge entomological club of a cast of the first paleozoic insect over found (p. 277), and of illustrations of the tertiary insects of North America (p. 278); as well as remarks on fossil species of Termes (p. 278); on some carboniferous insects, and on tertiary spiders from Florissant (p. 279).

—— Older fossil insects west of the Mississippi. (Proc. Bost. soc. nat. hist., 22: 53–60.) 8°. Boston. 1883.

Notice of Phlaeocoris, found in Missouri, and of the discovery of triassic cockroaches in Colorado.

—— New genera and species of fossil cockroaches, from the older American rocks. (Proc. acad. nat. sci. Philad., 1885: 34–39.) 8°. Philadelphia. 1885.

Description of four genera and eleven species from the triassic (8 species) or carboniferous (3 species) rocks.

Scudder, S. H.—Continued.

—— The geological history of myriapods and arachnids. (Psyche, 4 : 245–250.) 8°. Cambridge. 1885.

Much the same as the general remarks in Zittel's Handb. d. paläont. See also de Borre, A. P. de.

—— Systematische übersicht der fossilen myriopoden, arachnoideen, und insekten. (Zittel, Handb. d. palaeont., 1. abth., Palaeoz., 11: 721–831, figs. 894 1109.) 8°. München. 1885.

The first comprehensive review of these groups since the work of Pictet, in 1846, and Giebel, in 1856. Figures are given for nearly every family treated.

Translation.—Myriapodes. Arachnides. Insectes. (Zittel, Traité de paléontologie. Traduit par le Dr. C. Barrois, 11. Palaeoz., Part 1: 720–833, figs. 911–1126.) 8°. Paris. 1887.

Translated by Mr. A. Six.

—— Systematic review of our present knowledge of fossil insects, including myriapods and arachnids. 8°. Washington. 1886. (Bull. U. S. Geol. Surv., vol. 5, No. 31.) 8°. Washington. 1886. 128 pp.

The original text furnished Dr. Zittel for his Handbuch, where the modern groups were more condensed. No illustrations, however, are given.

Abstract.—Fossil insects. (Journ. roy. micr. soc., 1887 : 682.) 8°. London. 1887.

—— The cockroach of the past. (Miall, Struct. cockroach, pp. 205–220, figs 119–125.) 8°. London. 1886.

A popular review of our present knowledge of extinct cockroaches, closing with a table of the 177 species known.

—— The work of a decade upon fossil insects, 1880–1889. (Psyche, 5: 287–295.) 4°. Cambridge. 1890.

A review of the advance that has been made in paleontology during the previous ten years, directing attention to the more important or interesting papers.

—— The fossil insects of North America, with notes on some European species. 2 vols. 4°. New York. 1890. vol. 1, pp. 10, 455, pl. 35;—vol. 2, pp. (2), 734, pl. 28.

A collection of the author's quarto publications on fossil insects. The separate volumes have independent title pages and are elsewhere noted, vol. 1, on pretertiary insects, in the next entry, vol. 2, on tertiary insects, in Section VI.

—— The pretertiary insects of North America, including critical remarks on and descriptions of some Eu-

Scudder, S. H.—Continued.
ropean forms. 4°. New York. 1890. pp.
10, 455, pl. 1-7, 7a, 8-34.

Forms Vol. 1 of the author's Fossil Insects of
North America, with notes on some European
species. Includes all the papers on the older insects, published in quarto form by the author between 1806 and 1890, together with a final chapter
of a bibliographical nature. The separate papers,
excepting the last, are elsewhere noticed.

——— See also Marcou, J. B.,
**Packard, A. S., Trouessart, E.: White,
C. A.**

**Selys-Longchamps, Michel Edmond
Baron de, et Hagen, Hermann August.**
Revue des odonates ou libellules d'Europe. 8°. Bruxelles, etc., 1850, pp. 22,
408, pl. 11, tableaux 6.

Contains, pp. 356-364, Énumération des odonates
fossiles d'Europe, by Dr. Hagen, with a few notes
by Baron de Selys; and pp. 365-368, Note sur l'énumération des odonates fossiles d'Europe, by de
Selys. In the former, 39 species are enumerated
with synonymy and brief notes; in the latter
they are discussed by formations, and the conclusion reached that Æschnidæ preceded the Agrionidæ and Libellulidæ.

Sendel, Nathaniel. See Guérin-Méneville, F. E.

Serres, Pierre Marcel Toussaint de.
See Pictet de la Rive, F. G.

Six, A. See Scudder, S. H.

Spener, Christian Maximilian. See
Vallisneri, A.

Steizel, J. T. Ueber fossile arachnoideen. (Ber. naturwiss. gesellsch. Chemnitz, 10: 63.) 8°. Chemnitz, 1887.

Abstract of remarks on the geological development of Arachnida.

Swinton, A. H. Notes on certain fossil Orthoptera claiming affinity with the
genus Gryllacris. 8°. [London. 1874.]
pp. 5, pl. (Geol. mag., (2), 1: 337-341, pl.
14.) 8°. London. 1874.

Entitled on cover of separate: On fossil Orthoptera. Claims to show that the carboniferous
Corydalis brongniarti is a Gryllacris, and discusses the tertiary species which have been referred to Gryllacris.

——— Insect variety: its propagation
and distribution; treating of the colours,
dances, colours, and music in all grasshoppers, cicadæ, and moths; beetles,
leaf insects, bees, and butterflies; bugs,
flies, and ephemeræ; and exhibiting the
bearing of the science of entomology

Swinton, A. H.—Continued.
on geology. 8°. London, etc. [1880.] pp.
10, 326, pl. 7.

Notices the stridulation of extinct insects, pp.
163-164; and reviews the strata containing insect
remains in a discursive manner, pp. 260-271.

Taylor, J. E. The geological antiquity
of flowers and insects. (Pop. science review, 17 (n. s., 2): 36-52, figs.) 8°. London. 1873.

Points out "a broad parallelism between the
appearances of the more differentiated types of
the vegetable kingdom and the development or
appearance of various orders of insects;" and on
pp. 43-44 gives an account of the general distribution of insects in geological times.

Taylor, Richard Cowling. Illustrations of antediluvian zoölogy. Articulated animals. (Lond. Mag. nat. hist.,
3: 361.) 8°. London. 1830.

Notices Coleoptera from Stonesfield slate, coal
shale of Yorkshire, peat beds of Norfolk, Yorkshire, and Lincolnshire coasts, in a submarine
forest at Mount's Bay, and at the Danby coal
pits, Yorkshire.

Thorell, Tamerlan. On European spiders; part 1. Review of the European
genera of spiders, preceded by some observations on zoological nomenclature.
4°. Upsala. 1869-70. pp. 24, 242.

Contains remarks on fossil spiders, pp. 220-223,
in which the affinities of the described species
are discussed, and some new genera are founded.

Trouessart, E. Revue de paléontologie pour l'année 1887, dirigée par H.
Douvillé. Arthropodes: Insectes, Myriapodes et Arachnides. (Ann. géol.
univ., 4: 719-731). 8° Paris. 1888.
Principally occupied with an analysis of Scudder's contribution to Zittel's Handbuch.

Vallisneri, Antonio. Istoria del cameleonte affricano, e di varj animali
d'Italia. 4°. Venezia. 1715. pp. (8),
200, tav. 5, 8.

Contains, pp. 181-190, an Epistola to Vallisneri
by Spener upon various fossils, including insects,
which are mentioned from the rocks of Thüringen
on pp. 186-187, and from amber on p. 187. In the
latter he recounts as in his possession "muscas,
culices, araneas, formicas volantes, scolopendras
aliaque unicolenta."

Vidal y Careta, Francisco. Los insectos y las plantas; discurso. 8°. Habana. 1888. pp. [4], 28, 1 chart.

A general review of fossil insects, in geological
sequence, and their correlation with vegetable
life.

Vogt, Karl. Lehrbuch der geologie
und petrofactenkunde. Zum gebrauch

Vogt, K.—Continued.
bei vorlesungen und zum selbstunterrichte. Zweite vermehrte und gänzlich umgearbeitete auflage. 2 v. 8°. Braunschweig. 1851. Vol. 1, pp. 29, 642, pl. 8, (6);—vol. 2, pp. 31, 672, pl. (2).

Insects are meagerly treated, with two or three figures copied from other works, 1, pp. 336, 482, 638-939; 2, pp. 450, 451, 509-511.

Vollmar. Ueber fossile entomologie. (Gistl, Faunus, 2: 56-62.) 8°. München. 1835.

Of a general nature, closing with a list of fossil Coleoptera, borrowed from Keferstein.

Walchner, Fritz Hermann. Der practische naturforscher. Ein unentbehrliches hand- und hülfsbuch für freunde der naturwissenschaften. 8°. Karlsruhe. 1842-'44. pp. 1198. Each part also contains sep. t. p., and a f. table.

Abtheilung III, Der petrefactolog (1843), contains a chapter on Fossiler Insecten. pp. 534-539, in which, especially in foot-notes, the genera then known are enumerated; nothing new is added.

Walckenaer, Charles Athanase, baron, et **Gervais, Paul.** Histoire naturelle des insectes. Aptères. 4 vol. and atlas of 52 plates. 8°. Paris. 1837-1847; tom. 1 (1837), 2 t. p., pp. 6, 682;—tom. 2 (1837), 2 t. p., pp. 549;—tom. 3 (1844), 2 t. p., pp. 8, 476;—tom. 4 (1847), 2t. p., pp. 16, 623.

References to fossils, all at second hand, will be found in vol. 3, pp. 6, 70-72, 81, 128, 283, 449; and vol. 4, pp. 329-330, 345, 356, 360. Atlas fossilis from amber is described in vol. 1, p. 462.

Wallace, Alfred Russel. The geographical distribution of animals; with a study of the relations of living and extinct faunas as elucidating the past changes of the earth's surface. 2 v. 8°. London. 1876.—The same: New York. 2 v. 8°. 1876. Vol. 1, pp. 24, 503, pl. 13, maps 5;—vol. 2, pp. 10, 607, pl. 7, maps 2.

In a section entitled: Antiquity of the genera of insects, 1, pp. 166-168 (both editions), he concludes that "many of the larger and more important genera of insects date back to the beginning of the tertiary period, or perhaps beyond it; but the family types are far older." The section abounds in errors.

———— Colour in nature. (Nature, 19: 501-505.) 4°. London. 1879.

Review of Grant Allen's Colour sense, in which, p. 501, he contends for the probability of flowering plants and accompanying butterflies in the Carboniferous.

Wallerius, Johan Gottskalk. Systema mineralogicum, quo corpora mineralia in classes, ordines, genera et species suis cum varietatibus divisa describuntur, atque observationibus, experimentis et figuris aeneis illustrantur. Editio altera correcta. 2 vol. 8°. Vindobonae. 1778. Vol. 1, pp. (16), 442, (35), port, pl.;—vol. 2, pp. (12), 640, (60), pl.

§ 154, 2, pp. 536-546: Entomolithi, contains, under the heads of Typolithi and Entoma, a catalogue of the fossil insects then known.

Waterhouse, Charles O. See **Owen, R.**

Weyenbergh, H. Een kort overzigt der entomologische fossiele schatten van Teyler. (Tijdschr. entom., (2), 3: 195-196.) 8°. 'sGravenhage. 1867.

Brief notice of the fossil arthropods of Teyler's museum.

White, Adam. See **Hagen, H. A.**

White, Charles Abiathar. Progress of invertebrate paleontology in the United States for the year 1880. (Amer. nat., 15: 273-279.) 8°. Philadelphia. 1881.

Notices papers by Scudder.

———— and **Nicholson, H. A.** Bibliography of North American invertebrate paleontology, being a report upon the publications that have hitherto been made upon the invertebrate paleontology of North America, including the West Indies and Greenland. 8°. Washington. 1878. pp. 132. (Misc. publ. U. S. geol. surv. terr., 10.)

Includes descriptive notes to most of the entries.

———— Supplement to the Bibliography of North American invertebrate paleontology. (Bull. U. S. geol. surv. terr., 5: 143-152.) 8°. Washington. 1879.

Literature of 1878 with omissions from the previous list.

Winkler, T. C. Musée Teyler. Catalogue systématique de la collection paléontologique. 8°. Harlem. 1863. t. p., pp. 4, 608.

Arachnides, p. 421; Insectes, pp. 422-429. 265 numbers are given, including over two hundred undetermined species.

Woodward, Henry, et al. Notes on fossil insect remains. (Geol. mag., 10: 1-2.) 8°. London. 1873.

A résumé, by the editors of the Journal, of papers that have appeared on the subject in their magazine, by Phillips, Kirkby, Scudder, Woodward, and Butler.

Zittel, Karl A. See **Scudder, S. H.**

II.—GENERAL FOR PALEOZOIC TIME.

. See also under Section I.

Archiac, Étienne Jules Adolph Victo ricomte d' et **Verneuil,** Philippe Édonard de. On the fossils of the older deposits in the Rhenish provinces; preceded by a general survey of the fauna of the palezoic rocks, and followed by a tabular list of the organic remains of the Devonian system in Europe. (Trans. geol. soc. Lond., (2), vol. 6, pp. 303–410.) 4°. London. 1842.

Contains, p. 330, a section (ix) of a single paragraph on what was then known of palezoic insects.

Binney, Edward William. On two remarkable fossil insects from the lower coal measures near Huddersfield. (Proc. lit. phil. soc. Manch., 6: 59.) 8°. Manchester. 1867.

(Geol. mag., 4: 132.) 8°. London. 1867.

Notices the occurrence of Xylobius sigillariae and of a supposed coleopterous larva.

Borre, Alfred Prendhomme de. Note sur des empreintes d'insectes fossiles, découvertes dans les schistes houillers des environs de Mons. (Comptes rendus soc. ent. Belg., (2), xii : 4–7 : and discussions on same by Breyer and others, 7–8.) 8°. Bruxelles. 1875.

Describes and discusses the affinities of two fossil insects which he considers orthopterous; and of a third which he compares to a carboniferous Termes. Breyer considers one of the first wings as lepidopterous.

——— Sur trois nouveaux insectes fossiles. (Comptes-rendus soc. ent. Belg. (2), xviii : 17.) 8°. Bruxelles. 1875.

(Ann. soc. ent. Belg., 18, comptes rendus, 115.) 8°. Bruxelles. 1875.

Notice of the discovery of two Neuroptera from the same beds as Breyeria and of a supposed dipteron from the Jurassic beds of Luxembourg; M. de Borre informs me that closer examination proves the latter to be an homopteron.

——— Complément de la note sur des empreintes d'insectes fossiles. (Comptes rendus soc. ent. Belg. (2), xiii : 7–11; and discussion on same by Fologne, Lafontaine, Plateau, Breyer, and de Selys, 11–12.) 8°. Bruxelles. 1875.

Now maintains the correctness of Breyer's belief that one of the wings is lepidopterous, or prolepidopterous, and changes the generic name from Pachytylopsis, formerly given, to Breyeria. Fologne and Lafontaine contend that there are two

Borre, A. P. de—Continued.
overlapping wings. Plateau at first thought it the tip of a coleopterous wing of gigantic size, but withdrew his opinion. Breyer maintained the latter view impossible, and de Selys thought it rash to refer a reticulated wing to the Lepidoptera. These two papers, without the discussion, were republished separately as follows:

——— Notes sur des empreintes d'insectes fossiles découvertes dans les schistes houillers des environs de Mons. 8°. Bruxelles. 1875. pp. 1–10, pl. 5–6. Ann. soc. ent. Belg., 18, pp. 39–42, 56–60, pl. 5–6. 8°. Bruxelles. 1875.

Première note, pp. 1–6 (39–42). Seconde note, pp. 6–10 (56–60). Second note reprinted as follows:

Empreintes d'insectes fossiles découvertes dans les schistes des environs de Mons. (Journ. zool., 4: 291–97.) 8°. Paris. 1875.

Unaccompanied by the plate. Gervais adds brief notes.

——— Analyse de deux travaux récents de MM. Scudder et Ch. Brongniart sur les articulés fossiles. 8°. Bruxelles. 1885. pp. 7 (Comptes rend. soc. entom. Belg., (3), lxv: 131–137.) 8°. Bruxelles. 1885.

See also notes on same (Ibid, (3), lxviii: 67; (3), lxx: 63). 8°. Bruxelles. 1886. Deals with the general classification of paleozoic hexapods.

——— See also Scudder, S. H.; Gab es schon u. s. w.

Bradley, Frank Howe. Geology of Grundy County. (Geol. surv. Ill., 4, chapt. 13, pp. 190–206.) 8°. [Springfield.] 1870.

Contains p. 196, a list of the carboniferous insects of Mazon Creek.

Brauer, Friedrich. Systematisch-zoologische studien. 8°. Wien. 1885. pp. 177, pl. 1. (Sitzungsb. kais. akad. wiss. Wien, 91: 237–413, pl.) 8°. Wien. 1885.

Discusses in several places the relationship of many of the older insects, and especially of Engereon, with remarks on synthetic types.

——— Ansichten über die paläozoischen insecten und deren deutung. (Ann. naturh. hofmus., 1: 87–126, pl. 7–8.) 8°. Wien. 1886.

Expands his previously expressed views at greater length, reviewing in detail the structure of many of the older types, and strenuously opposing the views of Scudder upon their general classification.

Breyer, Albert. See Borre, A. P. de.

Brodie, P. B. Fossil insects in the carboniferous rocks. (Geol. mag., 4: 245-246.) 8°. London. 1867.

A brief enumeration of the different forms known to the author.

Brongniart, C. J. E. Sur un nouvel insecte fossile des terrains carbonifères de Commentry (Allier) et sur la faune entomologique du terrain houiller. (Bull. soc. géol. France, (3), 11: 142-151, pl. 4.) 8°. Paris. 1883.

An account of the insects of Commentry, with a description of Titanophasma, which is also figured, and ending with a general account and classed list of carboniferous insects.

——— Aperçu sur les insectes fossiles en général et observations sur quelques insectes des terrains houillers de Commentry (Allier, France). 16°. Paris. [1883.] 8 pp. (Le naturaliste, 5: 266-269.) 4°. Paris. 1883.

A brief general review, followed by an account of the rich locality of Commentry, largely based on his paper of the previous year read before the Geological society.

The same. ([Publications] soc. industr. minérale. Distr. du centre.) 8°. Montluçon. 1883. 15 pp., pl.

The same as preceding, with slight additions, especially in the description of Titanophasma, which is also figured, the description taken from his Geological society's paper.

——— Les insectes fossiles des terrains primaires. Comp d'œil rapide sur la faune entomologique des terrains paléozoïques. 8°. Rouen. 1885. [22] pp., [5] pl. (Bull. soc. amis sc. nat. Rouen, 1885: 50-68, pl. 1-3.) 8°. Rouen. 1885.

Forms part of the Compte rendu de la 23e réunion de délégués des sociétés savantes à la Sorbonne (1884) par Henri Gadeau de Kerville.

The paper is principally concerned with the carboniferous insects of Commentry, for which a large number of family, generic, and specific names are proposed. The separate copy contains two (unnumbered) plates not in the society's issue, and a sheet of errata. Many of the Commentry insects are figured on these plates for the first time. (See also Borre, A. P. de.)

TRANSLATION: Die fossilen insecten der primären schichten. (Jahrb. k. k. géol. reichsanst., 35: 649-662. 8°. Wien. 1885.

Contains no plates.

TRANSLATION: The fossil insects of the primary group of rocks: a rapid survey of the entomological fauna of the

Brongniart, C. J. E.—Continued. palæozoic systems. Transl. by Mark Stirrup. 12°. Salford. 1885. pp. 20. (Trans. Manch. geol. soc., 18: 269-288, pl.) 8°. Manchester. 1885.

A discussion, participated in by Messrs. Stirrup, Ormerod, Dickinson, Burnett, Boyd Dawkins, Wild, and Watts, will be found on pp. 267-269, 289-292, 328-331. No plate appeared in the separate.

Reprint of the last. (Geol. mag., n. s. (3), 2: 481-491, pl. 12.) 8°. London. 1885.

The plate is the same as plate 4 of the original, and is wrongly said to have been published by the Rouen Society.

ANALYSIS: (Entom. nachr., xi: 330-332.) 8°. Berlin. 1885.

ANALYSIS: (Proc. ent. soc. Lond., 1886: 3-8.) 8°. London. 1886.

By Herbert Goss.

ABSTRACT: Fossile insecten der primärzeit. (Kosmos, 19: 64-67.) 8°. Stuttgart. 1886.

ABSTRACT: Les insectes fossiles des terrains primaires. (Rev. scient., (3), 36: 275-278, fig.) 4°. Paris. 1885.

——— Coup d'œil rapide sur la faune entomologique des terrains paléozoïques. (Ann. géol. univ. 5: 1019-1024.) 8°. Paris. 1889.

The same in substance as the earlier paper with a similar title, but with some modifications due to subsequent researches.

Burnett, R. T. [Discussion of Stirrup's translation of Brongniart's paper.] (Trans. geol. soc. Manch., 18: 290.) 8°. Manchester. 1885.

Very general remarks.

Dawkins, William Boyd. [Discussion of Stirrup's translation of Brongniart's paper.] (Trans. geol. soc. Manch., 18: 328-329, 331.) 8°. Manchester. 1886.

Myriapods found in nodules in the valley of the Irwell near Clifton: Trigonocarpa bored by insects.

Dawson, Sir J. W. On the conditions of the deposition of coal, more especially as illustrated by the coal-formation of Nova Scotia and New Brunswick. (Quart. journ. geol. soc. Lond., 22: 95-169, pl. 5-12.) 8°. London. 1866.

Merely refers (p. 145) to the occurrence of a myriapod and one insect at the Joggins.

——— On some remains of palæozoic insects recently discovered in Nova Scotia and New Brunswick. (Can. nat. [n. s.],

Dawson, *Sir* J. W.—Continued.
3: 202–206, 5 woodc. in text.) 8°. Montreal. 1867.

(Geol. mag., 4 : 385–388, pl. 17, figs. 1–5.) 8°. London. 1867.

Haplophlebium barnesii and four of the Devonian insects are described and figured for the first time by Scudder.

ABSTRACT : On insects from the Carboniferous and Devonian formations. (Geol. mag., 4 : 374.) 8°. London. 1867.

Taken from the Montreal Gazette of May 1, 1867. An abstract also appears in Amer. Journ. sc., (2), 44 : 116. 8°. New Haven. 1867.

—— Acadian geology. The geological structure, organic remains, and mineral resources of Nova Scotia, New Brunswick, and Prince Edward Island. 2d edition, revised and enlarged, with a geological map and numerous illustrations. 8°. London. 1868. pp. 27, 694, pl. (9), map, figs. 231, (1), in text.

Pages 386–388, 524–526, figs. 151, 181–184, contain descriptions and illustrations of carboniferous and devonian insects by Scudder. There is also a Note on the myriapoda of the coal formation on pp. 495–496, by the same.

—— Supplement to the second edition of Acadian geology, containing additional facts as to the geological structure, fossil remains, and mineral resources of Nova Scotia, New Brunswick, and Prince Edward Island. 8°. London. 1878. pp. 102.

This supplement bound with reissue of 2d ed. forms 3d ed. Mentions and figures, pp. 53, 55, 56, some carboniferous insects and myriapods which had been described by Scudder since the previous edition.

—— Ancient insects and scorpions. (Can. rec. sc., 1 : 207–208.) 8°. Montreal. 1885.

Notice of Palæophonus nuncius and Palæoblattina douvillei.

Dickinson, Joseph. [Discussion of Stirrup's translation of Brongniart's paper.] (Trans. geol. soc. Manch., 18 : 289, 291, 328, 329.) 8°. Manchester. 1885–'86.

Believes all the paleozoic insects to be amphibians.

Dohrn, Anton. Zur kenntniss der insecten in den primärformationen. (Palaeontogr., 16 : 129–134, taf. 8.) 4°. Cassel. 1867.

Further discussion of Eugereon and description of two new Carboniferous insects. It is proposed

Dohrn, A.—Continued.
to extend the new order Dictyoptera, so as to embrace a number of the earlier insects.

—— Eugereon boeckingi und die genealogie der arthropoden. (Stett. entom. zeit., 28 : 145–153, pl. 1 [41].) 8°. Stettin. 1867.

Fuller discussion of the affinities of Eugereon and its bearings on Haeckel's views of the genealogy of insects.

Fleck, Hugo. See **Geinitz,** H. B., **Fleck,** H., und **Hartig,** E.

Fletcher, John. A dreadful phenomenon described and improved; being a particular account of the sudden stoppage of the river Severn, and of the terrible desolation that happened to the hirches between Coalbrook Dale and Buildwas Bridge in Shropshire on Thursday morning, May 27, 1773. (Works of John Fletcher, Vicar of Madeley, 1 : 229–246.) 12°. London. [1773?]

On p. 237 "a great many [fossils] were found bearing the impression of a dying insect, not unlike the butterfly into which silk-worms are changed."

Fologne, Egide. See **Borre,** A. P. de.

Fossil insects. (Amer. nat., 2 : 164, figs. 1, 2.) 8°. Salem. 1868.

Note (not original) on Xenoneura and Palæocampa.

Fritsch, A. Fauna der steinkohlenformation Böhmens. (Archiv naturw. landesdurchf. Böhmen, bd. 2, abth. 2, th. 1, pp. 1–16, pl. 1–4.) 8°. Prag. 1874.

Describes Palæranæa borassifolia for the first time, and gives new figures and descriptions of four previously-known insects, including the famous scorpion described by Corda.

—— Fauna der gaskohle und der kalksteine der permformation Böhmens. Bd. 1, heft 1. 4°. Prag. 1879. pp. 92, taf. 12.

Contains pp. 26–31: Vorläufige uebersicht der in der gaskohle und den kalksteinen der perforation in Böhmen vorgefundenen thierreste. On p. 31 appears a list of five insects, to three of which (Myriapoda) names are given, from Nýřan and Kounová.

Gab es schon während der steinkohlenzeit schmetterlinge? (Kosmos, 5 : 218–219.) 8°. Leipzig. 1879.

An account of the discussion of this subject, by Wallace, MacLachlan, de Borre, etc., in Nature and elsewhere.

Geinitz, Hans Bruno, **Fleck,** Hugo, und **Hartig,** Ernst. Die steinkohlen

Geinitz, H. B., etc.—Continued.

Deutschland's und anderer länder Europa's, ihre natur, lagerungs-verhältnisse, verbreitung, geschichte, statistik und technische verwendung. 2bd. 4°. München. 1865. Bd. 1 (*also entitled:* Geologie der steinkohlen Deutschland's und anderer länder Europa's, mit hinblick auf ihre technische verwendung; von Geinitz). pp. 10, 420, atlas, ff. 3, pl. 28. Bd. 2 (also entitled: Geschichte, statistik und technik der steinkohlen Deutschland's und anderer länder Europa's; von Fleck u. Hartig). pp. 8, 423, (4), map.

Contains (bd. 1, pp. 146-160), Organische überreste der steinkohlenformation des Saarbrückenschen, in which, pp. 149, 150, appear lists of the Carboniferous and Dyas insects of the basin of the Saar.

——— und Gutbier, A. von. Die versteinerungen von Obersaechsen und der Lausitz. (Gein., Gäa von Sachsen. pp. 61-142.) 8°. Dresden und Leipzig. 1843.

Insects at pp. 60, 116, 140; nothing new.

Germar, Ernst Friedrich. Beschreibung einiger neuen fossilen insecten (i) in den lithographischen schiefern von Bayern und (ii) in schieferthon des steinkohlengebirges von Wettin. (Münst., Beitr. z. petref., heft 5, pp. 79-94, taf. 9, 13.) 4°. Bayreuth. 1842.

The second part, pp. 90-94, pl. 13, describes and figures four cockroaches and a neuropteroid insect from the coal measures. See also Voigt, C. G. See also the same title in Section IV.

——— Die versteinerungen des steinkohlengebirges von Wettin und Löbejün in Saalkreise. *Also entitled:* Petrificata stratorum lithanthracum Wettini et Lobejani in circulo Salae reperta. f°. 8 hefte [fasc.]. Halle. 1844-53. pp. 4, 116, taf. [tab.], 40. 1es heft, pp. i.-iv., 1-12, pl. 1-5, 1844; 2es heft, pp. 15-28, pl. 6-10, 1845; 3es heft, pp. 29-40, pl. 11-15, 1845; 4es heft, pp. 41-48, pl. 16-20, 1847; 5es heft, pp. 49-60 (59), pl. 21-25, 1848; 6es heft, pp. 61-80, pl. 26-30, 1849; 7es heft, pp. 81-102, pl. 31-35, 1851; 8es heft, pp. 103-116, pl. 36-40, 1853.

Ueberreste von insekten [Insectorum vestigia], pp. 81-88, pl. 31, 39 (1851), almost entirely devoted to the cockroaches of the palæozoic rocks, on which it is the first important publication.

Gervais, Paul. See Borre, A. P. de.

Goldberger, F. See Goldenberg, F.

Goldenberg, Friedrich. Prodrom einer naturgeschichte der fossilen insecten der kohlenformation von Saarbrücken. (Sitzungsb. math.-nat. cl. akad. wiss. Wien, 9: 38-39.) 8°. Wien. 1852.

A nominal list, without description, of six new Orthoptera and Neuroptera. The author's name is accidentally given as Goldberger.

——— [Brief] an Herrn v. Carnell. (Zeitschr. deutsch. geol. gesellsch., 4: 246-248.) 8°. Berlin. 1852.

Much the same as the preceding, but with a few more details and comparisons, and without mention of specific names.

——— Ueber versteinerte Insectenreste im steinkohlengebirge von Saarbrücken. (Amtl. ber. vers. gesellsch. deutsch. naturf., 29: 123-126.) 4°. Wiesbaden. 1853.

Only the first half relates to insects in which a general account of his discoveries at Saarbrück is given; the latter half refers to the plants found in the same deposits.

——— Die fossilen insecten der kohlenformation von Saarbrücken. Cassel. 1854. t. p., pp. 24, pl. 4. 4°. (Palæontogr., 4: 17-40, tab. 3-6.) 4°. Cassel. 1854.

A careful description and excellent illustration of the species mentioned in his previous papers, with as many more. The remarkable genus Dictyoneura is introduced with three species.

——— Beiträge zur vorweltlichen fauna des steinkohlengebirges zu Saarbrücken. The title within is: Uebersicht der thierreste der kohlenformation von Saarbrücken. (Jahresb. k. gymn. u. vorsch. Saarbr., 1867, pp. 1-26.) 4°. Saarbrücken. 1867.

The insects occupy pp. 7-20 and swell the number of Saarbrück insects to seventeen. References are made to two plates, but these are not given until the same paper appears as the first heft of his Fauna saraep. foss.

——— Zur kenntniss der fossilen insecten in der steinkohlen-formation. (Neues jahrb. f. miner., 1869, pp. 158-168, pl. 3.) 8°. Stuttgart. 1869.

Description and illustration of ten new Blattinæ and two Homoptera.

——— Fauna saraepontana fossilis. Die fossilen thiere aus der steinkohlenformation von Saarbrücken. 1es heft, mit zwei tafeln abbildungen. 4°. Saarbrücken. 1873. t. p., pp. 26, (2), pl. 2. 2es heft, mit zwei tafeln abbildungen. 4°. Saarbrücken. 1877. pp. 4, 54, pl. 2.

The first part, with the exception of the introduction and the addition of the plates referred to

Goldenberg, F.—Continued.

In the text, is an exact reproduction of the paper published six years earlier in the report of the Saarbrück gymnasium, no mention being made of the author's own additions to the Carboniferous fauna since it was issued. These and others appear in the second part, where the insects occupy pp. 8-34 and pl. 1. The order Palæodictyoptera is here instituted for the Dictyoptera (nom. præocc.) of Dohrn. The number of species treated is twenty-seven, not, however, all confined to Saarbrück; this brings the number of Saarbrück insects as given in the catalogue, pp. 50-51, to thirty-eight, and renders this work, for its time, the most important contribution to palæozoic entomology that had ever appeared. A nominal list of 70 fossil cockroaches, based on that of Heer, is given on pp. 19-21. A supplement-heft was promised, but has never appeared.

Goss, H. Introductory papers on fossil entomology. No. 3. Palæozoic time. On the insects of the Devonian period, and the animals and plants with which they were correlated. (Entom. monthl. mag., 15: 124-127.) 8°. London. 1878.

The same.—No. 4. Palæozoic time. On the insects of the Carboniferous period, and the animals and plants with which they were correlated. (Entom. monthl. mag., 15: 169-173.) 8°. London. 1879.

The same.—No. 5. Palæozoic time. On the insects of the Permian period, and the animals and plants with which they were correlated. (Entom. monthl. mag., 15: 226-228.) 8°. London. 1879.

See the same title in Section I, Section IV, and Section VI.

―――― Three papers on fossil insects, and the British and foreign formations in which insect remains have been detected. No. 3. The insect fauna of the primary or palæozoic period. 8°. [London. 1880.] pp. 32. (Proc. geol. assoc., 6, no. 6, pp. 271-300.) 8°. London. 1880.

ABSTRACT: The insect fauna of the primary or palæozoic period and the British and foreign strata of that period in which insect remains have been detected. (Geol. mag. (n. s.), 6: 230-232.) 8°. London. 1879.

The papers, of which this is the third, contain a careful review of the literature of fossil insects; each geological formation is separately treated, containing references to all the genera, and in very many cases to the species found in it, with full bibliographical references. It will be found very useful to the general student.

The other papers are given in Section IV and Section VI, q. v.

Goss, H.—Continued.

―――― On some recently discovered insects from carboniferous and Silurian rocks. 8°. [London.] 1885. pp. [2], 21. (Proc. geol. assoc., 9, No. 3: 131-151.) 8°. London. 1885.

A general résumé of the progress of discovery since his previous collation, in which the deposits of Great Britain, the continent of Europe, and America are separately treated, followed by a general summary.

ABSTRACT. Fossil insects. (Entom., 18: 196-197.) 8°. London, 1885.

TRANSLATION OF ABSTRACT. Fossile insekten. Stett. ent. zeit., 46: 380-381.) 16°. Stettin. 1885.

Gutbier, August von. See Geinitz, H. B., und Gutbier, A. von.

Hartig, Ernst. See Geinitz, H. B., Fleck, H., und Hartig, E.

Higgins, H. H. President's address [to the Liverpool naturalists' field club]. 8°. [Liverpool. 1871.] 11 pp. 2 pl.

Mention, p. 10, of a few insect remains found in carboniferous rocks in the Ravenhead cutting near Liverpool; one wing is figured.

Humbert, Alois. See Scudder, S. H.

James, Joseph Francis. Remarks on a supposed fossil fungus from the coal measures. (Journ. Cinc. soc. nat. hist., 3: 157-159.) 8. Cincinnati. 1885.

Considers Rhizomorpha sigillariæ the burrow of an insect larva.

Kliver, Moritz. Ueber einige neue Blattinarien-, zwei Dictyoneura- und zwei Arthropleura-arten aus der Saarbrücker steinkohlenformation. 4°. Cassel. 1883. 19 pp., 3 pl. (Palæontogr., 29 (3,v): 249-264, pl. 34-36.) 4°. Cassel. 1883.

The Blattinariæ are seven in number, of which all but two are new, as are the species of Dictyoneura; all are figured.

―――― Ueber einige neue arthropodenreste aus der saarbrücker und der wettin-löbejüner steinkohlenformation. (Palæontogr., 32: 99-115, pl. 14 (7). 4°. Stuttgart. 1886.

Eleven different objects are described and figured, including some uncertain fragments of bodies; most are wings and include two Termes, a Dictyoneura, an Acridites and three Blattinariæ.

K[ušta, Johann]. Noví členovci z českého útvaru kamenouhelného. Vesmír 13: 97-98. (figs.) 4°. Praze. 1884.

Kušta, J.—Continued.

Popular account of the older insects of Bohemia with a list of a dozen species and figures of four or five.

—— Neue fossile arthropoden aus dem nöggerathienschiefer von Rakonitz. 8°. Prag. 1885. 8 pp. fig. (Sitzungsb. k. böhm. gesellsch. wiss., 1885: 592–597). 8°. Prag. 1885.

Describes Eolycosa and Eojulus and gives a list of seventeen species of arthropods from the carboniferous beds of Rakonitz.

Lacoe, R. D. List of palæozoic fossil insects of the United States and Canada, alphabetically arranged, giving names of authors, geological age, locality of occurrence, and place of preservation, with references to the principal bibliography of the subject. 8°. Wilkes-Barre, Pa. 1883. 21 pp. Publ. Wyom. hist. geol. soc., No. 5.)

Catalogues forty genera and seventy-two species.

Lafontaine, Jules de. See Borre, A. P. de.

McLachlan, R. See Gab es schon u. s. w.

Matthew, George F. On some remarkable organisms of the Silurian and Devonian rocks in southern New Brunswick. (Trans. roy. soc. Canada, 1888, sect. iv: 49–62, pl. 4.) 4°. Montreal. 1889.

Describes the wing of Geroneura wilsoni from the Devonian of Lancaster and a supposed larva, Archæoscolex, from the Devonian of St. John; a short section on the "Geological age of the insect remains" refers them to the middle Devonian.

Meek, Fielding Bradford, and Worthen, Amos Henry. Preliminary notice of a scorpion, a Eurypterus, and other fossils, from the coal-measures of Illinois. (Amer. journ. sc. arts, [2], 46: 19–28.) 8°. New Haven. 1868.

Afterwards described more fully in the Geological survey of Illinois. Among the "other fossils" are two myriapods.

—— Articulate fossils of the coal measures. (Worthen, Geol. surv. Ill., 3, ii, Palæontology, pp. 540–565, figs.) 8°. [Springfield.] 1868.

Describe and figure, pp. 556–565, two myriapods and two arachnids, with a Note on the genus Palæocampa, p. 565, first described as a caterpillar, but here considered a worm.

Miller, Samuel Almond. The American palæozoic fossils: a catalogue of the

Miller, S. A.—Continued.

genera and species, with names of authors, dates, places of publication, groups of rocks in which found, and the etymology and signification of the words, and an introduction devoted to the stratigraphical geology of the palæozoic rocks. 8°. Cincinnati. 1877. pp. 15, 253.

Arachnida, Myriapoda, and Insecta, pp. 224–226.

—— North American geology and palæontology for the use of amateurs, students, and scientists. 8°. Cincinnati. 1889. pp. 664, figs. 1194.

The Arachnida occupy pp. 569–571 with 3 figures; the Myriapoda pp. 572–574 with 6 figs.; and the Insecta pp. 574–581 with 23 figures, not all rightly named. Only palæozoic species are considered and it is not complete for those.

Morris, William. Did flowers exist during the Carboniferous epoch? (Nature, 20:404.) 8°. London. 1879.

Notices a "carbonaceous impress on a piece of shale from the Slievardagh coal field, Tipperary," which "appears" to him "to be a butterfly."

Murchison, R. I. The Silurian system, founded on geological researches in the counties of Salop, Hereford, Radnor, Montgomery, Caermarthen, Brecon, Pembroke, Monmouth, Gloucester, Worcester, and Stafford; with descriptions of the coal fields and overlying formations. 4°. London. 1839. pp. 32, 768, pl. 37.

—— Siluria. The history of the oldest known rocks containing organic remains, with a brief sketch of the distribution of gold over the earth. 8°. London. 1854. pp. 16, 523, pl. 37.

Notices insects of the coal, p. 284.

—— Siluria. The history of the oldest fossiliferous rocks and their foundations; with a brief sketch of the distribution of gold over the earth. 3d edition (including the Silurian system) with maps and many additional illustrations. 8°. London. 1859. pp. 20, 592, (2), pl. (2), 41, maps 2.

Brief reference to and figure of an insect "allied to Corydalis" from Coalbrookdale on pp. 320–321.

Noeggerath, Jacob. Die insecten der steinkohlenflora. (Frankf. convers.-blatt, 1851:215–216, 223.) 4°. Frankfurt a. M. 1851. (Kölnische zeitung, 1851.)

Not seen; quoted from Hagen's Bibliography.

Oldest (The) air breathers. (Pop. sc. monthl., 27: 395–400, 5 figs.) 8°. New York. 1885.

Popular account of Palæophonus and Palæoblattina, with extracts from the original accounts.

Ormerod, H. M. [Discussion of Stirrup's translation of Brongniart's paper.] (Trans. geol. soc. Manch., 18: 2–9, 291, 328.) 8°. Manchester. 1885, 1886.

The Commentry insects are preserved beside leaves.

Pike, J. W. Preservation of fossil insects and plants on Mazon Creek. 8°. Salem. 1881. pp. 5. (Proc. Amer. assoc. adv. sc., 29: 520–524.) 8°. Salem. 1881.

A sketch of the history of the formations at Mazon Creek, with a mention in most general terms of the animals and plants.

Plateau, Félix. See Borre, A. P. de.

Prestwich, Joseph. On some of the faults which affect the coalfield of Coalbrookdale. (Lond. Edinb. phil. mag., 4: 375–376.) 8°. London. 1834. (Proc. geol. soc. Lond., 2: 18–20.) 8°. London. 1834.

In a notice of the fossils a beetle and a spider are mentioned, p. 376 (20), as occurring in the iron-stone nodules at that place.

—— On the geology of Coalbrook Dale. (Trans. geol. soc. Lond., (2), 5:413–495, pl. 30–41.) 4°. London. 1840.

Contains a notice, p. 416, of three fossil insects (Curculioides [sic] ansticii, C. prestvicii, and a neuropteron) from the locality; also entered in the table on p. 490.

Scudder, S. H. [Notice of fossil insects from the Devonian rocks of New Brunswick, and on Haplophlebium barnesii.] (Proc. Bost. soc. nat. hist., 11: 150–151.) 8°. Boston. 1867. (Amer. nat., 1: 445–446.) 8°. Salem. 1867.

Refers the Devonian insects to new families of Neuroptera and the Carboniferous Haplophlebium probably to the Ephemeridæ.

—— [Remarks on two new fossil insects from the Carboniferous formation in America.] (Proc. Bost. soc. nat. hist. 11: 301–403.) 8°. Boston. 1868. (Scudder, Entom. notes, 1: 7–9.) 8°. Boston. 1868. (Amer. journ. sc., (2), 46. 419–421.) 8°. New Haven. 1868.

Describes Megathentomum pustulatum and Archegogryllus priscus. Abstract in Amer. nat., 2: 300. 8°. Salem. 1869.

Scudder, S. H.—Continued.

—— Entomological notes. I. 8°. [Boston. 1868.] pp. 11.

Reprint of the preceding paper, with others.

—— Descriptions of fossil insects, found on Mazon Creek and near Morris, Grundy Co., Illinois. (Worthen, Geol. surv. Ill., 3: 566–572, figs. 1–10.) 8°. [Springfield.] 1868.

Describes nine new Carboniferous insects, mostly neuropteroid. It is also marked as a Supplement to Descriptions of Articulates.

—— Remarks on some remains of insects occurring in Carboniferous shale at Cape Breton. (Proc. Bost. soc. nat. hist., 18: 113–114.) 8°. Boston. 1875. (Scudder, Ent. notes, 5: 2–3.) 8°. Boston. 1876.

Notices the discovery of cockroach wings and the larva of a supposed dragon-fly at Sidney.

—— New and interesting insects from the Carboniferous of Cape Breton. 8°. Salem. 1876. pp. 2, figs. (Proc. Amer. assoc. adv. sc., 24: B, 110–111, figs. 1–2.) 8°. Salem. 1876.

Description of the cockroach and supposed larval dragon-fly mentioned in a preceding paper.

REPRINT: 8°. Montreal. 1876. pp. 2, figs. (Can. nat., (n. s.), 8: 88–90, figs. 1–2.) 8°. Montreal. 1876.

—— Fossil palæozoic insects, with a list of described American insects from the Carboniferous formation. (Geol. mag., (n. s.), dec. 2, vol. 3, pp. 519–520.) 8°. London. 1876.

Gives a list of thirty species. Entitled in table: On fossil insects from the coal measures.

—— On the close affiliation of the insects of Europe and America in the Carboniferous epoch. (Proc. Bost. soc. nat. hist., 18:358–359.) 8°. Boston. 1876.

As intimately related as now.

—— Entomological notes, VI. 8°. Boston. 1878. pp. 55, pl.

Reprint, among other matter, of three short papers on fossil insects elsewhere mentioned.

—— The early types of insects; or the origin and sequence of insect life in palæozoic times (Mem. Bost. soc. nat. hist., 3: 13–24.) 4°. Boston. 1879.

Forms pp. 33–41 of the Fossil insects of North America, vol. 1.

A general review of palæozoic insects, attempting to show "that the laws of succession of the insect tribes are quite similar to those which have long been known to hold in other groups of the

Scudder, S. H.—Continued.

nnimal kingdom; and that the facts are, in the main, such as the theory of descent demands." The general conclusions are summarized under twelve heads. It is noticed and criticized by de Borre (Compt. rend. soc. ent. Belg., (2), no. 65, p. 11.) 8°. Bruxelles. 1879.

TRANSLATION: Les premiers types d'insectes; origine et ordre de succession des insectes dans la période paléozoïque. (Arch. sc. phys. nat., (3), 3: 353-371.) 8°. Genève. 1880.

The translation is by A. Humbert. Some notes, especially the bibliographical, are omitted.

ABSTRACT: The early types of insects. Abstract of a paper read before the National Academy of Sciences, Nov. 5, 1878. (Amer. journ. sc. arts, (3), 17: 72-74.) 8°. New Haven. 1879. (Science news, 1: 22-23.) 8°. Salem. 1878.

This contains the general conclusions.

TRANSLATION: Urtypen der insecten. (Kosmos, 5: 61-62.) 8°. Leipzig. 1879.

A translation of the abstract, with notes by the editor.

———— Two new British Carboniferous insects, with remarks on those already known. (Geol. mag., (2), 8: 293-300, fig.) 8°. London. 1881.

The only two hexapods hitherto known (excepting a cockroach) are Neuroptera and not Orthoptera as had been recently maintained. Two new species are described, Archæoptilus ingens, the largest palæozoic insect known, and Brodia priscotincta, remarkable for the preservation of the colored bands of the wing. Separates (without change of pagination) bear on the cover the title: New Carboniferous insects.

ABSTRACT: Upon the Carboniferous insects of Great Britain. (Harv. univ. bull., 2: 175.) 4°. Cambridge. 1881.

———— The Carboniferous hexapod insects of Great Britain. (Mem. Bost. soc. nat. hist., 3: 213-224, pl. 12.) 4°. Boston. 1883.

Forms vol. 1, pp. 235-246, pl. 11, of the Fossil Insects of North America.

Brodia, Archæoptilus and Lithosialis are described at length; a list of 7 species is given.

———— On additional remains of articulates obtained by Dr. Dawson from sigillarian stumps in the coal field of Nova Scotia. (Phil. trans., 1882: 649-650.) 4°. London. 1883.

Notices some new myriapods and scorpions without giving names.

Scudder, S. H.—Continued.

———— Palæodictyoptera: or the affinities and classification of paleozoic Hexapoda. (Mem. Bost. soc. nat. hist., 3: 319-351, pl. 29-32.) 4°. Boston. 1885.

Forms vol. 1, pp. 283-315, pl. 15-18, of the Fossil Insects of North America.

The winged insects of palæozoic times are grouped under the terms orthopteroid, neuropteroid and hemipteroid Palæodictyoptera, and separated into several families. A large number of new genera and species are described and figured from American rocks.

———— Winged insects from a paleontological point of view, or The geological history of insects. (Mem. Bost. soc. nat. hist., 3: 353-352.) 4°. Boston. 1885.

Forms vol. 1, pp. 317-322, of the Fossil Insects of North America.

The division of insects into two great groups, Metabola and Heterometabola, is urged upon new grounds; but ordinal features were not differentiated in the earliest period. "Insects continued through palæozoic time as a generalized form of Heterometabola . . . which had the front wings as well as the hind wings membranous. On the advent of mesozoic times a great differentiation took place, and before its middle all of the orders [of insects] . . . were fully developed in all their essential features as they exist to-day."

———— Two more English Carboniferous insects. (Geol. mag., (3), 2: 265-266.) 8°. London. 1885.

Brodia priscotincta and Archæoptilus ingens described.

———— See also Borre, A. P. de, Brauer, F., Dawson, J. W.

Selys-Longchamps, M. E. de. See Borre, A. P. de.

Sterzel, J. T. Ueber zwei neue insektenarten aus dem karbon von Lugau. (Ber. naturw. gesellsch. Chemnitz, 7: 271-276, pl.) 8°. Chemnitz. 1881.

Describes and figures Blattina (Etoblattina) lanceolata and Termes (Mixotermes?) lugauensis.

Stirrup, Mark. [Discussion of Brongniart's paper.] (Trans. geol. soc. Manch., 18: 267-269, 289, 292, 329-331.) 8°. Manchester. 1885, 1886.

Explanation of the conditions under which insects were found at Commentry: the excellence of their preservation; perforations in fossil woods in Lancashire.

———— See also Brongniart, C. J. E.

Verneuil, Philippe Édouard Poulletier de. See Archiac, E. J. A. V. d', et Verneuil, P. É. P. de.

Voigt [C. G.?] Neueste acquisitionen des Halle'schen minoralogischen museums. (Ber. naturw. ver. Harzes, 1840-46 (2° anfl.), p. 26.) 4°. Wornigerode. 1856.

Exhibition of four species of Blattina from Wettin and Löbejun, and of an insect to be called Acridites carbonatus by Germar; these were afterwards described by Germar in Münster's Beiträge.

Wallace, A. R. See Gab es schon n. s. w.

Watts, William. [Discussion of Stirrup's translation of Brongniart's paper.] (Trans. geol. soc. Manch., 18: 291.) 8°. Manchester. 1885.

Discusses the amphibious nature of the Commentry fossils.

Wild, George. [Discussion of Stirrup's translation of Brongniart's paper.] (Trans. geol. soc. Manch., 18: 290, 291.) 8°. Manchester. 1885.

Expresses his belief that there were insects in Carboniferous times; fossils from Oldham show insect borings.

Woodward, H. Notes on some fossil

Woodward, H.—Continued.

crustacea, and a chilognathous myriapod, from the coal measures of the west of Scotland. (Trans. geol. soc. Glasgow, 2: 234–248, pl. 3.) 8°. Glasgow. 1-67.

Describes a Xylobius on pp. 235-237, and enumerates the true insects from the coal formation on pp. 237-240.

ABSTRACT: Notes on a chilognathous myriapod and some fossil crustacea from the coal measures of the west of Scotland. (Geol. mag., 4: 130–131.) 8°. London. 1867.

—— On some supposed fossil remains of Arachnida (?) and Myriopoda from the English coal-measures. (Geol. mag., 10: 104–112, figs.) 8°. London. 1873.

Separate also entitled: On British fossil arthropoda. 8°. London. 1873. pp. 9, figs. 11. Discusses the affinities of "Eurypterus? (Euphoberia) ferox," referring it to the myriapods.

Worthen, Amos Henry. See Meek, F. B., and Worthen, A. H.

III.—SPECIAL FOR PALEOZOIC TIME.

IIIa.—Paleozoic Myriapoda.

. See also under Sections I and II.

Borre, A. P. de. Sur un travail récent de M. S. H.-Scudder concernant les myriapodes du terrain houiller. 8°. Bruxelles. 1882. pp. 3. (Comptes rend. soc. entom. belg., (3), xix: 103–105.) 8°. Bruxelles. 1882.

An analysis of his Archipolypoda, also entitled: Analyse d'un mémoire... sur les archipolypodes, nouvel ordre de myriapodes fossiles.

—— Tentamen catalogi Lysiopetalidarum, Julidarum, Archiulidarum, Polyzonidarum atque Siphonophoridarum hucusque descriptarum. 8°. Bruxelles. 1884. pp. 41. (Ann. soc. entom. Belg., 28: 46-82.) 8°. Bruxelles. 1884.

Six fossil species of Archiulidæ are catalogued on p. 38 (79).

Dawson, J. W. On a terrestrial mollusk, a chilognathous myriapod and some new species of reptiles from the coal for-

Dawson, J. W.—Continued.

mation of Nova Scotia. (Quart. journ. geol. soc. Lond., 16, i: 268–277, figs. 1–29.) 8°. London. 1859.

Describes and figures Xylobius sigillariæ.

—— The air breathers of the coal period in Nova Scotia. (Can. nat. geol., 8: 1-12, 81-92, 159-175, 268-295, pl. 1-6.) 8°. Montreal. 1863.

Same as the following.

—— Air breathers of the coal period: a descriptive account of the remains of land animals found in the coal formation of Nova Scotia, with remarks on their bearing on theories of the formation of coal and of the origin of species, with illustrations. 8°. Montreal. 1863. t. p., front., pp. 4, 81, pl. 6, and a plate of photogr.

Contains, section xii, Invertebrate air breathers, pp. 62-63, and pl. 6 (pars) which describes Xylobius sigillariæ. See also p. 67.

Dohrn, A. Julus brassii n. sp. ein myriapode aus der steinkohlenformation

Dohrn, A.—Continued.

(with note by Weiss). (Verhandl. naturh. ver. preuss. Rheinl. u. Westph., (3), 5: 335-336, pl. 6.) 8°. Bonn. 1868.

Description of a species from Lobach with memoranda of previously described species. The note by Weiss is purely geological, on the probable equivalence of the Lebach beds and those yielding Xylobius.

Geinitz, H. B. Paläontologische mittheilungen aus dem mineralogischen museum in Dresden. (Sitzungsb. naturw. gesellsch. Isis, 1872: 125-131, taf. 1.) 8°. Dresden. 1872.

Contains, pp. 128-131, taf. 1, fig. 4-7: Iii. Fossile myriapoden in dem rothliegenden bei Chemnitz. Palæojulus dyadicus is described.

————— Ueber Palæojulus dyadicus. (Neues jahrb. f. miner., 1873: 733.) 8°. Stuttgart. 1873.

In response to Sterzel, defends the myriapodan character of Palæojulus. The identity of his Palæojulus with Scolecopteris elegans Zenk. is acknowledged by the author in 1880. See his Nachträge zur Dyas, I. (Mitth. k. min. geol. prähist. mus. Dresden, heft. 3: 1-4). 4°. Cassel. 1880. See also Sterzel, J. T.

Heathcote, F. G. The post-embryonic development of Julus terrestris. (Phil. trans. roy. soc. Lond., 179, B: 157-179, pls. 27-30.) 4°. London. 1889.

Heathcote points out that the relations of the dorsal and ventral regions of the body of the young Julus correspond exactly with their permanent condition in Euphoberia, a Carboniferous myriapod; and he further holds that the traces of the division of the dorsal plates found in the Archipolypoda lend additional strength to the belief that they are composed in modern diplopods of two fused segments originally distinct; which the doubling of the internal organs and of the mesoblastic segmentation also indicates.

ABSTRACT: (Proc. roy. soc. Lond., 43: 243-245.) 8°. London. 1888.

Latzel, Robert. Die myriopoden der österreichisch-ungarischen monarchie. 2 v. 8°. Wien. 1880, 1884. Vol. 1, pp. 15, 228, pl. 10; vol. 2, pp. 12, 414, pl. 16.

Contains Einiges über fossile myriopoden, vol. 2, pp. 364-366. Based on Scudder's memoir on the Archipolypoda, followed by a list of fossil genera.

Meek, F. B., and Worthen, A. H. Notice of some new types of organic remains from the coal measures of Illinois. (Proc. acad. nat. sc. Philad., 1865; 41-53.) 8°. Philadelphia. 1865.

Meek, F. B., and Worthen, A. H.—Continued.

Describe Anthracorpes typus as a myriapod and Palaeocampa anthrax as a lepidopterous larva, both afterwards considered by them as worms. See also Worthen, A. H.

Packard, A. S. The systematic position of the Archipolypoda, a group of fossil myriapods. (Amer. nat., 17: 326-329, figs.) 8°. Philadelphia. 1883.

Considers that these Carboniferous myriapods should not be separated from the Chilognatha, but that they should form a group under the Chilognatha distinct from modern types.

————— On the morphology of the Myriopoda. (Proc. Amer. phil. soc., 21: 197-209, figs.) 8°. Philadelphia. 1883.

At the close discusses the probable affinities of Palaeocampa, which he regards as probably a neuropterous larva.

Page, David. Advanced text book of geology, descriptive and industrial. 12°. Edinburgh. 1860. pp. 10, 326, figs.

On p. 135 figures Kampecaris.

Peach, Benjamin N. On some fossil myriapods from the lower old red sandstone of Forfarshire. 8°. Edinburgh. 1882. 11 pp., pl. (Proc. roy. phys. soc. Edinb., 7: 177-188, pl. 2.) 8°. Edinburgh. 1882.

Describes the genera Kampecaris and Archidesmus with one species of each.

Salter, John William. On some species of Eurypterus and allied forms. (Quart. journ. geol. soc. Lond., 19, i: 81-87, figs.) 8°. London. 1863.

Describes and figures as a Eurypterus some Carboniferous fragments since recognized as myriapods.

Scudder, S. H. On the fossil myriapods of the coal formations of Nova Scotia and England. (Quart. journ. geol. soc. Lond., 25, i: 441.) 8°. London. 1869.

Briefest abstract of the next paper published four years later.

————— On the Carboniferous myriapods preserved in the sigillarian stumps of Nova Scotia. (Mem. Bost. soc. nat. hist., 2: 231-280.) 4°. Boston. 1873.

Forms pp. 21-29 of the Fossil insects of North America, vol. 1.

Describes two genera and several species besides the Xylobius sigillariae of Dawson Abstract in Amer. journ. sc., (3), 6: 225-226. 8°. N. Haven. 1873.

————— Supplementary note on fossil myriapods. 4°. Boston. 1878. p. 1, figs. (Mem. Bost. soc. nat. hist., 2: 561-562, figs.) 4°. Boston. 1878.

Scudder, S. H.—Continued.

Forms p. 31 of the Fossil insects of North America, vol. 1. Description of figures of the different species, omitted from the preceding paper.

———— The structure and affinities of Euphoberia Meek and Worthen, a genus of Carboniferous myriapoda. (Amer. journ. sc., arts, (3), 21: 182–186.) 8°. New Haven. 1881.

Points out the distinction between the Carboniferous and modern diplopodous myriapods, and proposes a distinct suborder, Archipolypoda, for the former.

———— [Exhibition of Carboniferous centipedes.] (Proc. Bost. soc. nat. hist., 21: 82.) 8°. Boston. 1881.

Archipolypoda from Illinois.

———— Remarks on remarkable Carboniferous millipedes. (Proc. Bost. soc. nat. hist., 21: 122.) 8°. Boston. 1881.

They form a distinct suborder.

———— Archipolypoda, a subordinal type of spined myriapods from the Carboniferous formation. (Mem. Bost. soc. nat. hist., 3: 143–182, pl. 10–13.) 4°. Boston. 1882.

Forms vol. 1, pp. 195–234, pl. 7*, 8–16, of the Fossil insects of North America.

Discusses their relation to modern diplopods and monographs the known species, adding several new ones; twelve species are recognized, divided into four genera. The first plate contains a restoration of the largest species with other animals and plants of Mazon Creek. See also Borre, A. P. de.

———— The affinities of Palaeocampa Meek and Worthen, as evidence of the wide diversity of type in the earliest known myriapods. (Amer. journ. sc., (3), 24: 161–170.) 8°. New Haven. 1882.

Critical examination of the structure of Palaeocampa formerly regarded as a lepidopterous larva or a worm, to show that it must be referred to the myriapods, but form an independent group therein.

———— Two new and diverse types of Carboniferous myriapods. (Mem. Bost. soc. nat. hist., 3: 283–297, pl. 26, 27, figs. 1–4.) 4°. Boston. 1884.

Forms vol. 1, pp. 247–261, pl. 12–13, of the Fossil insects of North America.

Trichiulus (afterwards discovered to be a folded fern-leaf), with three species, is described, and Palaeocampa, whose structure is described at length, is classed with myriapods as a new suborder, Protosyngnatha.

———— Note on the supposed myriapodan genus Trichiulus (Mem. Bost. soc. nat. hist., 3: 438.) 4°. Boston. 1886.

Scudder, S. H.—Continued.

Forms part of p. 330, vol. 1, of the Fossil insects of North America.

Trichiulus proves to be the tip of an unfolding fern-leaf.

———— New Carboniferous myriapoda from Illinois. (Mem. Bost. soc. nat. hist., 4: 417–442, pl. 33–38.) 4°. Boston. 1890.

Forms pp. 393–418, pl. 25–30, of the Fossil Insects of North America, vol. 1.

Sixteen species are described and three genera, besides notes on five species previously known.

———— See also Latzel, R.

Sterzel, J. T. Ueber Palaeojulus dyadicus. (Neues jahrb. miner., 1878: 729–731.) 8°. Stuttgart. 1878.

Considers the fossil described by Geinitz as a myriapod to be a fern-leaf, of a species of Scolecopteris.

———— Ueber Palaeojulus dyadicus Geinitz und Scolecopteris elegans Zenker. (Zeitschr. deutsch. geol. gesellsch., 30: 417–426, taf. 19.) 8°. Berlin. 1878.

Mainly devoted to showing that no animal remains have been found which possess the characters assigned to Palaeojulus dyadicus; but this so-called myriapod is the half of a leaf of Scolecopteris elegans, a fern.

———— Ueber Scolecopteris elegans Zenker und andere fossile reste aus dem hornstein von Altendorf bei Chemnitz. (Zeitschr. deutsch. geol. gesellsch., 32: 1–18, taf. 1–2.) 8°. Berlin. 1880.

Has a brief reference to Palaeojulus dyadicus on pp. 1–2.

Weiss, Christian Ernst. See **Dohrn, A.**

Westwood, John Obadiah. Sur une empreinte trouvée en Angleterre dans les schistes houillers et supposée pouvoir être celle d'une chenille. (Compt. rend. soc. ent. Belg., (2), xxii: 6–7.) 8°. Bruxelles. 1876. (Ann. soc. ent. Belg., 19, Compt. rend., 4–5.) 8°. Bruxelles. 1876.

Translation by Weyers of a passage of Westwood's from Brodie's work on the secondary insects of England, in which a supposed caterpillar (since recognized as a myriapod) is described. Weyers adds a few comments.

Weyers, Joseph Leopold. See **Westwood, J. O.**

Woodward, H. A monograph of the British fossil Crustacea belonging to the order Merostomata. 4°. London. 1866–1878. t. p., pp. 2, 263, pl. 36. ([Publ.] paleontogr. soc., 1866–78.)

Woodward, H.—Continued.

Includes, pp. 171-174, in part iv (1872), Eurypterus (Euphoberia) ferox, since held by Woodward and others to be a myriapod.

—— On Euphoberia brownii H.
Woodw., a new species of myriapod from
the coal-measures of the west of Scotland.
(Geol. mag., 8: 102-104, pl. 3, figs. 6-7.)
8°. London. 1871.

Describes and compares with E. armigera of the
Illinois Carboniferous beds.

—— On some spined myriapods from
the Carboniferous series of England.
(Geol. mag., n. s., dec. 3, vol. 4: 1-10,
figs., pl. 1.) 8°. London. 1887.

Separate under the title: Myriapods of the coal
period. Gives a historical summary of published
data concerning Carboniferous myriapods and discusses at length with the aid of figures the structure of Euphoberia ferox.

—— Supplementary note on Euphoberia ferox. (Geol. mag., n. s., dec. 3,
vol. 4: 116-117, figs.) 8°. London. 1887.

The illustration in Brodie's work is faulty, so
that E. ferox can not be referred to Acanther-
pestes.

Worthen, A. H. Geological survey of
Illinois. Vol. 2. Palæontology. 8°.
[Springfield] 1866. pp. 22, 6, (2), 11-170,
8 pl. 50.

In Section II, Descriptions of invertebrates
from the carboniferous system by F. B. Meek
and A. H. Worthen, occur on pp. 409-411 descriptions of Anthracorpes, a supposed myriapod, and Palæocampa, regarded as a caterpillar,
but since shown to be a myriapod. Both are figured on pl. 32.

—— See Meek, F. B., and Worthen,
A. H.

IIIb.—Paleozoic Arachnida.

. See also under Sections I and II.

Beecher, Charles E. Note on the fossil
spider Arthrolycosa antiqua Harger.
(Amer. journ. sc., (3), 38: 219-223, 3 figs.)
8°. New Haven. 1889.

"It seems necessary to exclude the genus from
the order Anthracomarti, and at present a strict
interpretation of any of the orders will not admit
this form."

Beneden, Edouard van. De la place
que les limules doivent occuper dans la
classification des arthropodes. (Comptes
rend. séances soc. ent. Belg., 1871-72: 9-
11.) 8°. Bruxelles. 1871. (Ann. soc.
ent. Belg., 15, compte rend., 9-11.) 8°.
Bruxelles. 1872.

Beneden, E. van—Continued.

TRANSLATION: On the systematic po-
sition of the king crabs and trilobites.
(Ann. mag. nat. hist., (4), 9: 98-99.) 8°.
London. 1872.

Concludes from a study of the embryonic development of Limulus that it presents the greatest
analogy to that of scorpions and other arachnids
from which the king crabs can not be separated.
"The Trilobites as well as the Eurypterida and
the Poecilopoda must be separated from the class
Crustacea, and form, with the Scorpionida and
the other Arachnida, a distinct branch, the origin
of which has still to be ascertained."

Brongniart, C. J. E. Palæophoneus
nuncius. (Bull. d. séances soc. ent.
France, 1884: 226.) 8°. Paris. 1884.
Brief notice.

Corda, August Karl Joseph. Ueber den
in der steinkohlenformation bei Chomle
gefundenen fossile scorpion. (Verhandl.
gesellsch. vaterl. mus. Böhm., vers. 13:
35-43, pl., figs. 1-14.) 8°. Prag. 1835.

Contains, a. Urtheil der commission bei der
naturforscherversammlung in Stuttgart, pp. 35-
36. b. Mikroskopische untersuchung, abbildung
und beschreibung, von Corda, pp. 36-43, with a detailed description and discussion of the affinities
of the scorpion.

—— Ueber eine fossile gattung der
afterscorpione. (Verhandl. gesellsch.
vaterl. mus. Böhm., vers. 17: 14-1c, pl. 1,
figs. 1-9.) 8°. Prag. 1839.

Description of Microlabis sternbergii. An abstract will be found in the Neues Jahrb. f. miner.,
1841: 854-855. 8°. Stuttgart. 1841.

Feistmantel, Karl. Die steinkohlen
becken in der umgebung von Radnic.
(Archiv naturw. landesdurchf. Böhmen,
bd. 1, sect. 2, at end, pp. 1-120, pl. 1-2.)
8°. Prag. 1869.

Gives, p. 66, a brief account of the Chomle scorpions described by Corda, and notices the discovery of Palæranea borassifolia without description.

Geikie, Archibald. A recent find in
British palæontology. (Nature, 25: 1-3.)
4°. London. 1881.

Contains an announcement of the discovery of
remarkably perfect scorpions in the coal measures
of Scotland, with some results of their study by
Mr. B. N. Peach. An abstract entitled New Carboniferous fossils in Scotland, extracted from the
London Times, will be found in Amer. nat., 16:
1021-1022. 8°. Philadelphia. 1881. See also Primeval Scottish scorpions.

Geinitz, H. B. Kreischoria wiedei
H. B. Gein., ein fossiler pseudoscorpion
aus der steinkohlenformation von

Geinitz, H. B.—Continued.

Zwickau. (Zeitschr. deutsch. geol. gesellsch., 1882: 238-242, pl. 14.) 8°. Berlin. 1882.

Besides a full description and figures of both dorsal and ventral surface comparisons are made with Architarbus and other allied forms.

Goss, H. Further evidence of the existence of insects in the Silurian period. (Ent. monthl. mag., 21 : 234.) 8°. London. 1885.

Mere reference to the discovery of the Scottish scorpion.

Harger, Oscar. Notice of a new fossil spider from the coal measures of Illinois. (Amer. journ. sc. arts, [3], 7 : 219-223, fig.) 8°. New Haven. 1874.

Extended description of Arthrolycosa antiqua with discussion of its affinities.

Harvey, F. L. On Anthracomartus trilobitus Scudd. (Proc. acad. nat. sc. Philad., 1886: 231-232.) 8°. Philadelphia. 1886.

Note on the geological position with a brief description of the fossil.

Haughton, Samuel. Description of a fossil spider, Architarbus subovalis, from the middle coal measures, Burnley, Lancaster. (Journ. geol. soc. Ireland, n. s., 1 : 222-223, figs.) 8°. Dublin, etc. 1877.

Gives a figure, without description, both of the original and of the new and better specimen.

Hunter, J. R. S. Notes on the discovery of a fossil scorpion (Palæophonus caledonicus) in the Silurian strata of Logan Water. (Trans. geol. soc. Glasg., 8 : 169-170.) 8°. Glasgow. 1886.

Brief statement regarding the manner and place of discovery with nothing of palæontological importance.

———— Notes on a new fossil scorpion (Palæophonus caledonicus) from the upper Silurian shales, Logan Water, Lesmahagow. (Trans. Edinb. geol. soc., 5 : 187-191, pl. 4.) 8°. Edinburgh. 1887.

Description with geological notes. Plate not published.

Karsch, F. Ueber ein neues spinnenthier aus der schlesischen steinkohle und die arachniden der steinkohlenformation überhaupt. (Zeitschr. deutsch. geol. gesellsch., 34 : 556-561, pl. 21.) 8°. Berlin. 1882.

Describes Anthracomartus and makes the first essay at a classification of palæozoic Arachnida.

Kinnear, W. T. On the occurrence and range of beds containing scorpion

Kinnear, W. T.—Continued.

remains in the Carboniferous rocks of Fife. (Trans. Edinb. geol. soc., 5 : 216-220.) 8°. Edinburgh. 1887.

Wholly geological, except for notes on the condition of specimens found.

Koninck, Laurent Guillaume de. Sur une nouvelle espèce de crustacé du terrain houiller de la Belgique. (Bull. acad. roy. Belg., (2), 45 : 409-410.) 16°. Bruxelles. 1878.

Introducing Woodward's communication on Brachypygo carbonis.

Kušta, J. Notiz über den fund eines arachnidenrestes im carbon bei Petrovic. 8°. Prag. 1882. 4 pp. (Sitzungsb. k. böhm. gesellsch. wiss., 1882: 258-261.) 8°. Prag. 1883.

A scorpion which he compares to Cyclophthalmus.

———— Anthracomartus krejčii, eine neue arachnide aus dem böhmischen carbon. 8°. Prag. 1883. 8 pp., pl. (Sitzungsb. k. bohm. gesellsch. wiss., 1883 : 340-345, pl.) 8°. Prag. 1883.

Extended descriptions and comparisons.

————. Ein neuer fundort von Cyclophthalmus senior C. 8°. Prag. 1884. 3 pp. (Sitzungsb. k. böhm. gesellsch. wiss., 1884 : 48-50.) 8°. Prag. 1884.

Description of a specimen found near Rakonitz.

———— Thelyphonus bohemicus n. sp. ein fossiler geisselscorpion aus der steinkohlenformation von Rakonitz. 8°. Prag. 1884. 8 pp., 2 pl. (Sitzungsb. k. böhm. gesellsch. wiss., 1884 : 186-191, 2 pl. in 1.) 8°. Prag. 1884.

Describes at length and figures excellently the first known Carboniferous pedipalp.

———— Neue arachniden aus der steinkohlenformation von Rakonitz. 8°. Prag. 1885. 8 pp., pl. (Sitzungsb. k. böhm. gesellsch. wiss., 1884 : 395-403, pl.) 8°. Prag. 1884.

Describes Rakovnicia and two species of Anthracomartus with additional remarks on Cyclophthalmus and Thelyphonus.

———— O nových arachnidech z karbonu Rakovnického. (Věstn. král. české spolecn. nauk, 1888: 192-208, pl.) 8°. Prag. 1888.

Describes Geralycosa, Scudderia, Entarbus, a new Anthracomartus, three species of Geralinura and an Anthrascorpio. A German abstract is added, with a list of arthropods from Rakonitz.

Lankester, Edwin Ray. Limulus an arachnid. (Quart. journ. micr. sc., 21: 504-548, 609-649, pl. 27-29, figs.) 8°. London. 1881.

An extended argument to show, mainly on anatomical and morphological grounds, that the king crabs are not Crustacea but Arachnida. With the Trilobita and Eurypterina he considers the Xiphosura as the precursors of the higher Arachnida.

—— Studies in Apus, Limulus, and Scorpio. 8°. London. 1882. t. p., pp. 36 (2), t. p., 8, 8, (4), pl. (3), figs.

Contains the same as the preceding with another paper.

Lindström, G. Sur un scorpion du terrain silurien de Suède. (Comptes rendus, 99 : 9-4-9-5.) 4°. Paris. 1884.

First announcement of Palaeophonus.

—— Le plus ancien animal terrestre connu. (La nature, 13, i : 33-34, fig.) 8°. Paris. 1884.

Brief account of Palaeophonus nuncius, based on a letter from Lindström to Milne Edwards, and accompanied by a woodcut.

—— List of the fossils of the upper Silurian formation of Gotland. 8°. Stockholm. 1885. 20 pp.

Palaeophonus placed at the head of the list.

—— Preliminary notice on a Silurian scorpion (Palaeophonus nuncius Thorell & Lindström), from the Isle of Gotland. (Trans. Edinb. geol. soc., 5: 151-156, pl. 3.) 8°. Edinburgh. 1887.

Brief description of the condition and organs of the specimen. The plate was not published.

—— See also **Thorell,** T., and Lindström, G.

Meyer, C. E. H. von. Scorpion aus dem steinkohlengebirg. (Mus. senckenb., 1 : 293.) 4°. Frankfurt a. M. 1834.

A simple notice of Sternberg's, or rather Corda's scorpion.

Mosely, H. N. See **Packard,** A. S.

Packard, A. S. Is Limulus an arachnid? (Amer. nat., 16 : 287-292.) 8°. Philadelphia. 1882.

In answer to Lankester, attempts to prove that Limulus is a crustacean, chiefly from the discovery by Willemoes of a nauplius-stage in the development of an East Indian species, a discovery which he afterwards acknowledges to be false. p. 436. See, also, Moseley in Nature, 25 : 582.

Peach, B. N. On some new species of fossil scorpions from the Carboniferous rocks of Scotland and the English borders, with a review of the genera Eoscor-

Peach, B. N.—Continued.

pius and Mazonia of Messrs. Meek and Worthen. (Trans. roy. soc. Edinb., 30 : 397-412, pl. 22, 23.) 4°. Edinburgh. 1882.

Description and illustration of five distinct forms referred to Eoscorpius, with a characterization of the group.

—— Ancient air breathers. (Nature, 31 : 295-298, 2 figs.) 8°. London. 1885.

General account of the Swedish and Scottish discoveries of Palaeophonus, with figures of both.

—— See also **Geikie,** A.

Primæval Scottish scorpions. (The Scotsman, no. 11960, p. 3.) f°. Edinburgh. 1881.

An anonymous communication, of more than a column in length, based on Geikie's article in Nature, with special reference to Peach's discoveries.

Reuss, A. E. Kurze übersicht der geognostischen verhältnisse Böhmens; fünf vorträge, gehalten im naturwissenschaftlichen vereine Lotos im jahre 1853; mit drei geologischen übersichtskarten. 8°. Prag. 1854. pp. 103, pl. 3.

Refers, p. 59, to two scorpions and a spider from the Carboniferous rocks of Bohemia.

Roemer, F. On a fossil spider lately discovered in the coal measures of Upper Silesia. (Report Brit. assoc. adv. sc., 1865, notices, 73.) 8°. London. 1866. (Geol. mag., 2 : 468.) 8°. London. 1865.

These two are essentially the same, with slightly differing titles; and are practically an abstract of the following.

—— Protolycosa anthracophila, eine fossile spinne aus dem steinkohlen-gebirge Oberschlesiens. (Neues jahrb. f. mineral., 1866 : 136-143, taf. 3.) 8°. Stuttgart. 1866.

Describes at length and figures Protolycosa anthracophila.

—— Auffindung und vorlegung eines neuen gliederthieres in dem steinkohlengebirge der Ferdinandsgrube bei Glatz. (Jahresh. schles. gesellsch. vaterl. cultur, 56 : 54-55.) 8°. Breslau. 1879.

Describes Arthitarbus sleslacus.

Scudder, S. H. A contribution to our knowledge of paleozoic Arachnida. (Proc. Amer. acad. arts sc., 20 : 13-22.) 8°. Boston. 1884.

A systematic revision of the known forms, with descriptions of many new genera and species.

Scudder, S. H.—Continued.

—— Note on Anthracomartus carbonis. 8°. Bruxelles. 1885. 2 pp., fig. (Compte rendu soc. ent. Belg., (3), lxii: 84–85, fig.) 8°. Bruxelles. 1885.

Woodward's Brachypyge carbonis is not a crustacean but an arachnid.

—— Illustrations of the carboniferous Arachnida of North America of the orders Anthracomarti and Pedipalpi. (Mem. Bost. soc. nat. hist., 4: 443–456, pl 39–40.) 4°. Boston. 1890

Forms pp. 419–432, pl. 31–32, of the fossil insects of North America, vol. 1. Ten species of eight genera are described and figured.

Sternberg, Kasper. Vortrag . . . in der allgemeinen versammlung des böhm. musenms am 14 April, 1835. (Verh. gesellsch. vaterl. mus. Böhm., 1835: 12–30.) 16°. Prag. 1835.

Contains, pp. 23–24, a history of the discovery of the scorpion described in samevolume by Corda.

Stur, D. Die culm-flora der ostrauer und waldenburger schichten. (Beiträge zur kenntniss der vorwelt, heft ii.) (Abhandl. k. k. geol. reichsanst., 8, ii: 1–14, 1–366, pl. A–C, 18–47.) 4°. Wien. 1877.

Figures und describes, p. 5, note, Euphrynus salml.

Thorell, T. On Proscorpius osbornei Whitfield. (Amer. nat., 20: 269–274.) 8°. Philadelphia. 1886.

Regards it as more nearly allied to Palaeophonus than to the Carboniferous scorpions, with which Whitfield had classed it.

—— and **Lindström, G.** On a silurian scorpion from Gotland. 4°. Stockholm. 1885. 33 pp., 2 pl. (Kongl. svensk. vetensk. akad. handl., 21, no. 9.) 4°. Stockholm. 1885.

Extended description and detailed illustration of Palaeophonus nuncius, followed by a classification of known fossil scorpions, and a characterization of the extinct groups.

Extract: (Biol. centralbl., 5: 657–661.) 8°. Erlangen. 1886.

Whitfield, Robert P. An American Silurian scorpion. (Science, 6: 87–88, fig.) 8°. New York. 1885.

First description and figure of Palaeophonus osborni.

—— On a fossil scorpion from the Silurian rocks of America. (Bull. Amer. nuns. nat. hist., 1: 181–190, pl. 19–20.) 8°. New York. 1885.

Whitfield, R. P.—Continued.

Full description and illustration of Proscorpius osborni, the first known Silurian scorpion in America.

—— Professor Thorell and the American Silurian scorpion. (Science, 7: 216–217.) 8°. New York. 1886.

A rejoinder to Thorell's criticism of the author's opinion of the systematic position of Proscorpius.

Woodward, H. On the discovery of a new and very perfect arachnide from the ironstone of the Dudley coal-field. (Geol. mag., 8: 385–388, pl. 11.) 8°. London. 1871.

Describes and figures a new specimen of the arachnid, to which Buckland, supposing it a beetle, formerly gave the name of Curculioides prestvichi. A list of fifty-four paleozoic insects is appended. An abstract will be found in Rep. Brit. assoc. adv. sc., 41, not., 112–113. 8°. London. 1872.

—— On a new arachnide from the coal measures of Lancashire. 8°. London. 1872. pp. 3. (Geol. mag., 9: 385–387.) 8°. London. 1872.

Describes Architarbus subovalis and compares it with A. rotundatus from the Carboniferous beds of Illinois. Also entitled on cover of separata: On a new fossil arachnide.

—— On the discovery of a fossil scorpion in the British coal measures. (Quart. journ. geol. soc. Lond., 32: 57–59, pl. 8.) 8°. London. 1876.

Describes and figures three distinct fragments of a scorpion, referred to one species called Eoscorpius anglicus. An abstract, under the title: The discovery of a fossil scorpion in the English coal measures is given in Hardw. sc. gossip, 1876: 20. 8°. London. 1876.

—— Discovery of the remains of a fossil crab (Decapoda-Brachyura) in the coal measures of the environs of Mons, Belgium. (Geol. mag., (2), 5: 433–436, pl. 11.) 8°. London. 1878.

Separate under the title: Discovery of a fossil crab in the coal measures of Belgium. 8°. London. 1878. pp. 4, pl. 11.

Description and illustration of Brachypyge carbonis, presumed to be the abdomen of a female brachyuran, but since shown to be that of an arachnid.

—— Découverte d'une espèce de décapode brachyure, dans le terrain houiller des environs de Mons. (Bull. acad. roy. Belg., (2), 45: 410–415, pl.) 16°. Bruxelles. 1878.

Much the same as the preceding, with the same plate. See also Koninck, L. G. de; and Scudder, S. H.

IIIc.—Neuropteroidea.

. See also under Section I and Section II.

Andree, Richard. Die versteinerungen der steinkohlenformation van Stradonitz in Böhmen. (Neues jahrb. f. mineral. geol. und pal., 1864, heft 2: 160–176, taf. 4.) 8°. Stuttgart. 1864.

Describes and figures Acridites priscus.

Assmann, A. See **Roemer, F.**

Audouin, [Jean] Victor. Sur une empreinte d'aile d'un insecte névroptère inconnu . . . trouvé en Angleterre à Colebroskedale [sic] dans le Shropshire . . . dans un terrain houiller. (Ann. soc. ent. France, 2, bull. ent., 7–8.) 8°. Paris. 1833.

The first mention of a palæozoic insect. It is said to have been shown by Audouin at the meeting of the Association of German naturalists at Bonn in 1835, but I have been unable to examine the report of that meeting. Also mentioned by Boué in his Résumé des progrès de la géologie, 1833, p. 146, and in the Journ. d. géol., 3: 105 (neither seen).

Beneden, P. J. van, et Coemans, Eugène. Un insecte et un gastéropode pulmoné du terrain houiller. 8°. pp. 20, pl. (Bull. acad. roy. Belg., (2), 23, iv: 384–401, pl.) 8°. Bruxelles. 1867.

Describe in detail and figure Onalia macroptera, which the authors consider allied to Hemeroblus.

REPRINT: Note sur un insecte et un gastéropode pulmoné du terrain houiller. (Ann. sc. nat., (5), zoöl., 7: 264–277, pl. 1, figs. 1–10.) 8°. Paris. 1867.

Borre, A. P. de. See **Volxem, C. van.**

Boué, Ami. See **Audouin, V.**

Bradley, F. H. Geology of Vermilion county. (Geol. surv. Ill., 4, chapt. 16, pp. 241–265.) 8°. [Springfield.] 1870.

Refers, p. 253, to a species of Miamia from the Carboniferous rocks near Georgetown.

Brongniart, C. J. E. La présence d'articulés dans les terrains siluriens. (Bull. d. séances soc. ent. France, 1884: 236–237.) 8°. Paris. 1884.

Brief notice of Palæoblattina douvillei.

—— [Dasyleptus lucasi.] (Bull. ent. soc. ent. France, 1885: 101–102.) 8°. Paris. 1885.

* A Carboniferous thysanuran.

TRANSLATION: Fossil thysanura. (Psyche, 4: 353.) 4°. Cambridge. 1890.

Brongniart, C. J. E.—Continued.

—— Corydaloides scudderi. (Bull. d. séances soc. ent. France, 1885: 12–13.) 8°. Paris. 1885.

Brief description of a Carboniferous insect from Commentry, remarkable for the lateral appendages of the abdomen.

—— Insecte fossile du terrain houiller. (La nature, 13, ii: 156, figs. 1, 2.) 8°. Paris. 1885.

The same as the last in popular form and with figures.

TRANSLATION: A new fossil insect. (Pop. sc. news, 20: 62, figs. 1, 2.) 4°. Boston. 1886.

—— *Sur la découverte d'une empreinte d'insecte dans les grès siluriens de Jurques (Calvados). (Comptes rendus, 99: 1164–1165.) 4°. Paris. 1885.

First description of Palæoblattina douvillei.

ABSTRACT: Insecte silurien. (La nature, 13, ii: 79.) 8°. Paris. 1885.

—— Insecte fossile des grès siluriens. (La nature, 13, ii: 116, figs.) 8°. Paris. 1885.

Virtually the same as the last paper in the Comptes rendus, but with figures.

Coemans, Eugène. See **Beneden, P. J. van, et Coemans, E.**

Cornelius, C. [Ueber termiten.] (Verh. naturh. ver. preuss. Rhein. Westph., 14: 20–44.) 8°. Bonn. 1857.

Contains, pp. 40–44, a letter from Dr. Hagen, criticising the treatment of the fossil Termites in Goldenberg's paper in Palæontographica.

Dana, J. D. On fossil insects from the Carboniferous formation in Illinois. (Amer. journ. sc. arts, (2), 37: 34–35, figs. 1–2.) 8°. New Haven. 1864.

Description and figures of two neuropteroid insects, the first recorded from the American coal formations.

Dawson, J. W. The oldest known insects. (Nature, 24: 483–484.) 4°. London. 1881.

Shows the Devonian age of the cordaite shales of New Brunswick, in which the then oldest known insects occur, from stratigraphical and botanical evidence, in opposition to the assumption of Hagen.

Dohrn, Carl August. Zur literatur über fossile Insecten. (Stett. entom. zeit., 42: 388–389.) 8°. Stettin. 1881.

Gives a résumé of Hagen's criticism of Scudder's Devonian insects.

Eaton, A. E. The oldest fossil insects. (Nature, 23:50.7) 4°. London. 1881.

Reply to Scudder's criticisms (in Devonian insects) of remarks on fossil Ephemeridæ in Eaton's Monograph.

Fritsch, A. Dva noví členovci z útvaru kamenouhelného v Čechách. (Vesmír, 9: 241-242, figs. A, B.) 4°. Praha. 1880.

Popular account of interesting remains of Ephemeridæ from the Carboniferous schists of Votvovic.

Geinitz, H. B. Ueber einige seltene versteluerungen aus der unteren dyas und der steinkohlen-formation. (Neues jahrb. f. miner., 1865: 385-391, taf. 2, 3.)

Describes Ephemerites rückerti with a letter upon the same from Dr. Hagen.

Goss, H. On the recent discovery of the wing of a cockroach in rocks belonging to the Silurian period. (Ent. monthl. mag., 21: 199-200.) 8°. London. 1885.

Notice of Palæoblattina douvillei.

TRANSLATION: Die jüngste entdeckung eines blattflügels in felsen, die zur silurischen periode gehören. (Stett. ent. zeit., 46: 134-136.) 16°. Stettin. 1885.

Haase, Erich. Die vorfähren der insecten. (Sitzungsb. gesellsch. Isis, 1886, abhandl., 85-91.) 8°. Dresden. 1886.

Makes reference to Palæoblattina, but has little to say on paleontological ground.

Hagen, H. A. The oldest fossil insects. (Nature, 23: 483-484.) 4°. London. 1881.

Disputes unqualifiedly the conclusions of Scudder concerning the Devonian insects.

—— The oldest fossil insects. (Nature, 24: 356-357.) 4°. London. 1881.

Is chiefly concerned with the determination of a species of Pecopteris found in the Devonian insect beds, and the conclusion is drawn that "those oldest insects will have to be considered as belonging to the lower Carboniferous." See also Dawson, J. W.

—— The Devonian insects of New Brunswick. (Bull. mus. comp. zoöl., 8: 275-284.) 8°. Cambridge. 1881.

After a "detailed comparison of a majority of the types," arrives at conclusions "radically different from the views entertained by Mr. Scudder," in his memoir on the subject.

—— Fossil insects. (Nature, 32: 53-54.) 4°. London. 1885.

Criticism of Scudder's paper of 1885 on Devonian insects.

Hagen, H. A.—Continued.

—— Die devonischen insecten. 8°. Leipzig. 1885. 5 pp. (Zool. anz., 8: 296-301.) 8°. Leipzig. 1885.

Discussion of the systematic position of the forms described by Scudder.

—— See also Dohrn, C. A.; Cornelius, C.; Geinitz, H. B.; Scudder, S. H.

Heer, O. Fossile flora der Bären insel. 4°. Stockholm. 1871. pp. 51, pl. 15. (Kongl. svensk. vetensk. akad. handl., 9, no. 5.)

In a section on the equivalents of the Bear Island beds in America, he refers the Devonian of St. John to the lower Carboniferous and mentions four species of insects described from there by Scudder.

Humbert, Alois. See Scudder, S. H.

Krause, Ernst. See Scudder, S. H.

McLachlan, R. See Woodward, H.

Novák, Ottomar. Ueber Gryllacris bohemica, einen neuen locustidenrest aus der steinkohlenformation von Stradonitz in Böhmen. 8°. Wien. 1880. pp. 6, pl. 1. (Jahrb. geol. reichsanst. Wien, 30: 69-74, taf. 2.) 1. 8°. Wien. 1880.

Compares the new species with Gryllacris brongniarti (Mantell), which also is figured, and to which it is closely allied. Reviewed by Scudder. (Proc. Bost. soc. nat. hist., 21: 167.) 8°. Boston. 1881.

Roemer, F. Vorlegung eines im schieferthon der zwischen Königshütte und Laurahütte gelegenen Alfredgrube 10 m. im liegenden des Carolinenflötzes gefundenen insectenflügels. (Jahresb. schles. gesellsch. vaterl. cultur, 62: 226.) 8°. Breslau. 1885.

Referred by Assmann to the vicinity of Gomphides.

Scudder, S. H. On the Devonian insects of New Brunswick. 8°. [Fredericton, 1865.] 1 p. (Bailey, Obs. geol. southern N. Brunsw., pp. 140-141.) 8°. Fredericton. 1865. (Amer. journ. sc., (2), 39: 357-358.) 8°. N. Haven. 1865. (Can. nat. geol., n. s., 2: 234-236.) 8°. Montreal. 1865. (Trans. entom. soc. Lond., (3), 2, proc., 117-118.) 8°. London. 1866.

Short statement of the probable affinities of the then oldest known insects, several species of which are distinguished.

—— On the fossil insects from Illinois, the Mhania and Hemeristia. (Amer. journ. sc., (2), 40: 269-271.) 8°. N. Haven. 1865.

Discussion of the affinities of two carboniferous insects, previously described by Dana.

Scudder, S. H.—Continued.

—— [Remarks on some fossil insects from the Carboniferous formation of Illinois and from the Devonian rocks of New Brunswick.] (Proc. Bost. soc. nat. hist., 10: 95–96.) 8°. Boston. 1865.

Merely an abstract of the two preceding papers.

—— An inquiry into the zoological relations of the first-discovered traces of fossil neuropterous insects in North America; with remarks on the difference in structure in the wings of living Neuroptera. 4°. [Boston.] 1866. pp. 20, pl. (Mem. Bost. soc. nat. hist., 1: 173-192, pl. 6.) 4°. Boston. 1866.

Forms pp. 1-20, pl. (1), of the Fossil insects of North America, vol 1.

Full treatment of the structure and relationship of Miamia and Hemeristia, which are referred to separate families, distinct from recent types. The wing structure of the modern families is also systematically reviewed.

—— An insect wing of extreme simplicity from the coal formation. (Proc. Bost. soc. nat. hist., 19: 243-249.) 8°. Boston. 1878. (Scudder, Entom. notes, 6: 37-38.) 8°. Boston. 1878.

Describes Euephemerites primordialis (which is probably a plant).

—— A Carboniferous termes from Illinois. (Proc. Bost. soc. nat. hist., 19: 300-301.) 8°. Boston. 1878. (Scudder, Entom. notes, 6: 54-55.) 8°. Boston. 1887.

Describes Termes continens.

—— The Devonian insects of New Brunswick. 4°. Boston. 1880. pp. 41, pl. 1. (Anniv. mem. Bost. soc. nat. hist., scient. papers, art. [3.]) 4°. Boston. 1880 [1881]. pp. 41, pl. 1.

Forms vol. 1, pp. 155-193, pl. 7, of the Fossil insects of North America.

Detailed description of six species, the oldest then known, with a study of their affinities. A jurassic may-fly is also described and the general conclusions summarized at the close. An extended analysis (by H. A. Hagen) is given in Zoöl. jahresb., 1880, II : 188-192. 8°. Leipsig. 1881.

EXTRACT: Relations of Devonian insects to later and existing types. (Amer. jour. sc. arts, (3), 21: 111-117.) 8°. New Haven. 1881. (Ann. mag. nat. hist., (5), 7: 255-261.) 8°. London. 1881.

Gives the general conclusions in full.

ABSTRACT: The Devonian insects. (Amer. nat., 14: 905-907.) 8°. Philadelphia. 1880.

Scudder, S. H.—Continued.

ABSTRACT: The Devonian insects of New Brunswick. (Science, 1: 292-293.) 4°. New York. 1880.

The above give the general conclusions somewhat condensed. More or less extended abstracts or notices will also be found in Arch. sc. phys. nat., (3),5 : 291-293 (by A. Humbert). 8°. Genève. 1881 ;—Bull. soc. ent. Ital., 12: 270-280. 8°. Firenze. 1881 ;—Naturforscher, 1881 : 141-143. 4°. Berlin. 1881 ;—Nation, 32, 150. 4°. New York, 1881. Also criticisms of the whole or part in Nature, 23 : 423-484 (by H. A. Hagen). 4°. London. 1881 ;—Bull. miss. comp. zool., 8: 275-284 (by the same). 8°. Cambridge. 1881 ;—and Nature, 23 : 507 (by A. E. Eaton). 4°. London. 1881.

TRANSLATION: Die devonischen Insekten von Neu-Braunschweig und ihre beziehungen zu den spätern und noch lebenden insekten. (Kosmos, 5 : 217-223.) 8°. Stuttgart. 1881.

A translation of the general conclusions, with an abstract of the other portions, and comments by the editor, Ernst Krause.

—— On Lithosialis bohemica. (Proc. Bost. soc. nat. hist., 21: 167.) 8°. Boston. 1881.

Reviews Novák's description of Gryllacris bohemica, showing that it is neuropterous not orthopterous.

—— The earliest winged insects of America. A reexamination of the Devonian insects of New Brunswick in the light of criticisms and of new studies of other paleozoic types. 4°. Cambridge. 1885. 8 pp. pl.

Forms vol. 1, pp. 275, 282, pl. 14, of the Fossil insects of North America.

A reply to the criticisms of Hagen.

—— See also Dawson, J. W.; Eaton, A. E.; Hagen, H. A.

Volxem, Camille van. Note critique sur l'interprétation donnée par M. de Borre d'une . . . empreinte fossile, celle du Pachytylopsis persenairei. (Comptes rendus soc. ent. Belg., (2), xxiv : 20-26, fig.) 8°. Bruxelles. 1876. (Ann. soc. ent. Belg., 19 : 23-34.) 8°. Bruxelles. 1876.

This posthumous note is introduced by explanatory remarks of Weyers. The author concludes that it is doubtful whether the fossil is the wing of an orthopteron; that it certainly is not an acridian, and has nothing in common with Pachytylus.

Also published separately entitled : Note critique sur le Pachytylopsis persenairei (de B.) par feu M. Camille Van Volxem. 8°. pp. 7.

Waterhouse, C. O. See Woodward, H.

Westwood, J. O. See Woodward, H.

Weyers, J. L. See Volxem, C. van.

Woodward, H. On a remarkable fossil orthopterous insect from the coal measures of Scotland. (Quart. journ. geol. soc. Lond., 32: 60-64, pl. 9.) 8°. London. 1876.

Describes and figures Lithomantis carbonarius, which with Westwood, Waterhouse, and McLachlan, he considers as belonging to "the neighborhood of the Mantidae." Corydalis (Gryllacris) brougniarti and a modern species of Blepharis are also figured for comparison as related to it. A list of fifty-eight paleozoic insects is appended. An abstract, under the title: Remarkable fossil orthopterous insect from the coal measures of Britain, appeared in Hardw. sc. gossip, 1876: 20. 8°. London. 1876.

IIId.—Orthopteroidea.

.*. See also under Section I and Section II.

Andrae, Karl Justus. Eine alge und eine insectenflügel aus der steinkohlenformation Belgiens. (Sitzungsb. niederrhein. gesellsch. Bonn, 1876: 27-28.) 8°. Bonn. 1877.

Mere mention of a Blattina.

Bar [Constant]. Sur la Breyeria borinensis. (Comptes rendus soc. ent. Belg., (2), xxviii : 6.) 8°. Bruxelles. 1876. (Ann. soc. ent. Belg., 19, comptes rendus, 53-54.) 8°. Bruxelles. 1876.

Considers Breyeria an homopterous insect, and maintains that the lack of flowers in Carboniferous times is not proof of the absence of Lepidoptera.

Borre, A. P. de. [Acknowledgment of objections to the lepidopterous character of Breyeria from Hagen, Heer, Giard, and Scudder.] (Comptes rendus soc. ent. Belg., (2), xxii : 5-6; followed by discussion of the subject by Saporta and Weyers, pp. 6-7.) 8°. Bruxelles. 1876. (Ann. soc. ent. Belg., 19 : 3-4.) 8°. Bruxelles. 1876.

—— Note sur le Breyeria borinensis. pp. 6. 8°. Bruxelles. 1879. (Comptes rendus soc. ent. Belg., (2), lxv : 7-12.) 8°. Bruxelles. 1879. (Ann. soc. entom. Belg., 22, compte rend., 77-83.) 8°. Bruxelles. 1879.

Maintains Breyeria to be an ancestral stock of the type of Lepidoptera.

—— See also Giard, A.

Brongniart, C. J. E. Note sur la découverte d'un orthoptère coureur de la famille des phasmiens (Protophasma dumasi) dans les terrains supra-houillers de Commentry (Allier). (Bull. séances soc. ent. France, 1878, vii : 69-70.) 8°. Paris. 1878.

The same with omission of the word "Note" in the title. (Ann. soc. ent. France, (5), 8, bull., 57-58.) 8°. Paris. 1878. (Bull. hebdom. assoc. sc. France, 22 : 172-173.) 8°. Paris. 1878.

The same as the last. 8°. Bruxelles. 1878. pp. 4. (Comptes rendus soc. ent. Belg., (2), xlvii : 9-12.) 8°. Bruxelles. 1878.

See the next entry.

—— Note sur un nouveau genre d'orthoptère fossile de la famille des phasmiens provenant des terrains suprahouillers de Commentry (Allier) (Protophasma dumasii). 8°. Paris. 1878. pp. 9, pl. (Ann. sc. nat., (6), zool., tom. 7, art. 1, pp. 8, pl. 6.) 8°. Paris. 1878.

Text substantially the same as the preceding.

A discovery carrying this group of Orthoptera back at once from the tertiary to the carboniferous. Noticed by Dr. Hector George in the Feuilleton of Le constitutionnel, Nov. 6, 1878 ; see also Girard, M.

TRANSLATION.—On a new genus of orthopterous insects of the family Phasmidae (Protophasma dumasii) from the upper coal measures of Commentry, dept. Allier, France. 8°. London. 1879. pp. 6, pl. The cover entitled : New fossil insect from the coal measures. (Geol. mag., n. s., dec. 2, vol. 6, pp. 97-102, pl. 4.) 8°. London. 1879.

—— Note sur les insectes fossiles de Commentry. (Compte rendu soc. géol. France, 1882-83 : 15-16.) 8°. Paris. 1883. (Bull. soc. géol. Fr., (3), 11 : 240-241.) 8°. Paris. 1883.

Brief description of Titanophasma.

—— [Sur le Titanophasma fayoli.] (Compte rendu soc. géol. France, 1882-83 : 23.) 8°. Paris. 1883.

Notice of a wing probably belonging to this Carboniferous insect.

—— Note complémentaire sur le Titanophasma fayoli et sur les Protophasma dumasii et woodwardii. 8°. Paris. 1883. pp. 2. (Bull. soc. entom. France, 1883, 20-21.) 8°. Paris. 1883.

Additions and corrections to former statements.

Brongniart, C. J. E.—Continued.

―― Un nouvel insecte fossile. (L'écho des mines et de la métallurgie viii; 5–6.) 4°. Lyon. 1883.

Much the same as his paper upon Titanophasma, with similar title, read before the Geological society in 1882, but with the omission of the list of Carboniferous insects.

―― Sur un gigantesque neurorthoptère, provenant des terrains houillers de Commentry (Allier). (Comptes rendus, 98: 832–833.) 4°. Paris. 1884.

Describes Titanophasma fayoli. Noticed in Ent. nachr., 10: 118 8°. Berlin. 1884.

―― Restaurations d'ailes d'insectes provenant du terrain carbonifère de Commentry (Allier). Bull. d. séances soc. ent. France, 1884: 225–226.) 8°. Paris. 1884.

Brief description of Dictyoneura ingens, with notice of the fauna of Commentry.

―― Les blattes de l'époque houillère. pp. 3. (Comptes rendus, 108: 252–254.) 4°. Paris. 1889.

Nearly six hundred cockroaches have been found in the Carboniferous deposits of Commentry, among which the Mylacridæ, not before recognized in the Old World, are as numerous as the Blattinariæ, and the structure of other parts of the body supports the division above mentioned, which had been based on characters drawn from the wings only.

Clark, Edgar F. Studies in the Rhode Island coal measures. (Proc. Newport nat. hist. soc., 2: 9–12.) 8°. Newport. 1884.

Notices discovery of a cockroach.

Dawson, J. W. Note on some new animal remains from the Carboniferous and Devonian of Canada. (Quart. journ. geol. soc. Lond., 26, i: 166.) 8°. London. 1870.

Notices the occurrence of cockroaches from the Nova Scotia coal measures.

Deichmüller, J. V. Ueber einiger blattiden aus dem brandschiefern der unteren dyas von Weissig bei Pillnitz. (Sitzungsb. gesellsch. Isis, 1882: 33–44, pl. 1.) 8°. Dresden. 1882.

Describes and figures two species of Etoblattina and one of Oryctoblattina.

―― Ueber zwei blattinen-reste aus den unteren lebacher schichten der Rheinprovinz. (Ber. senckenb. gesellsch. Frankfurt, 1886–'87: 89–94, pl. 3.) 8°. Frankfurt. 1887.

Describes two species of Etoblattina.

Deichmüller, J. V. See also **Geinitz, H. B.**

Eaton, A. E. Did flowers exist during the Carboniferous epoch? (Nature, 20: 315.) 4°. London. 1879.

Breyeria is an ephemerid.

TRANSLATION: Der angebliche steinkohlenzeit - schmetterling. (Kosmos, 5: 461–462.) 8°. Leipzig. 1879.

Elrod, Moses N., and **McIntire, E. S.** Report of a geological survey of Orange County. (Ann. rep. geol. surv. Ind., 7: 203–239.) 8°. Indianapolis. 1876.

The geological position of Paolia vetusta is shown on pp. 206, 221.

Fontaine, William Morris, and **White, I. C.** The Permian or upper Carboniferous flora of West Virginia and S. W. Pennsylvania. 8°. Harrisburg. 1880. (Rep. progr. second geol. surv. Penn., PP. pp. 10, 143, map, pl. 39.)

Contains a description, p. 104, and a figure, pl. 38, figs. 5, 5ᵃ, of Gerablattina balteata, by S. H. S[cudder].

Geinitz, F. E. Versteinerungen aus dem brandschiefer der unteren dyas von Weissig bei Pillnitz in Sachsen. 8°. Stuttgart. 1873. pp. 14, pl. (Neues jahrb. f. miner., 1873: 691–704, taf. 3.) 8°. Stuttgart. 1873.

Describes two Blattinæ. See same title in Section IIIᵃ.

―― Ueber neue aufschlüsse im brandschiefer der unteren dyas von Weissig bei Pillnitz in Sachsen. 8°. Stuttgart. 1875. pp. 14, pl. (Neues jahrb. f. miner., 1875, 1–14, taf. 1.) 8°. Stuttgart. 1875.

II. Insecten, pp. 4–6; describes four species of Blattina, one of them as new.

―― Die blattinen aus der unteren dyas von Weissig bei Pillnitz. 4°. Halle. 1880. pp. 22, pl. 1. (Nova acta k. leop.-carol.-deutschen akad. naturf., 41, ii, no. 7, pl. 39.) 4°. Halle. 1880.

Gives a full description, with figures, of the diversity in neuration in opposite wings, of a species of cockroach, together with criticisms of Scudder's Palæozoic cockroaches, and notes and figures of seven other forms.

Geinitz, H. B. Bericht über die ... auf dem reviere des carlschachtes der luganiederwitschnitzer steinkohlenwerke gesammelten steinkohlenpflanzen.) Sitzungsb. naturw. gesellsch. Isis, 1879, 7–13, taf. 1.) 8°. Dresden. 1879.

Geinitz, II. B.—Continued.

Describes, with Deichmüller, pp. 12-13 (two figures in text), Blattina dresdensis from the coal beds near Klein-Opitz, Saxony.

George, Hector. See **Brongniart, C. J. E.**

Giard, Alfred. Un papillon dans la houille; note do M. Prendhomme de Borre. (Bull. scient. hist. et litt. dép. Nord, 7 : 121-127.) 8°. Lille. 1875.

Discusses the affinities of Breyeria, concluding that it is not a lepidopteron but belongs to the Archiptera.

——— See also **Borre, A. P. de.**

Giebel, C. G. Ueber insectenreste im wettiner steinkohlengebirge. (Jahresb. naturw. ver. Halle, 2 : 8-9.) 8°. Berlin. 1850.

Mention of the cockroaches described by Germar.

Girard, M. Un spectre fossile. (La nature, 7 : 108-110, fig.) 4°. Paris, 1879.

Popular account of Protophasma dumasi.

Goeppert, Heinrich Robert. Die fossile flora der permischen formation. 4°. Cassel. 1864-65. 2 t. p., pp. 316, taf. 64. (Palaeontogr., bd. 12.) 4°. Cassel. 1864-65.

Contains, p. 289: D. Beiträge zur fauna der permischen formation, in which he mentions and names two wings and a body of insects regarded as cockroaches (localities not specified) which are figured on plates 28 and 64.

Goldenberg, F. Zwei neue ostracoden und eine blattina aus der steinkohlenformation von Saarbrücken. (Neues jahrb. f. mineral., 1870 : 286-289, figs.) 8°. Stuttgart. 1870.

Blattina winteriana is described and figured on pp. 288-289.

——— Beitrag zur insectenfauna der kohlenformation von Saarbrücken. (Zeitschr. ges. naturw., (3), 6 : 184-187, figs.) 8°. Berlin. 1881.

Describes Anthracoblattina scudderi.

Hagen, H. A. See **Borre, A. P. de.**

Heer, O. New orthopterous insect in the coal measures of Scotland. (Geol. mag., (2), 3 : 520.) 8°. London. 1876.

Notes some omissions of fossil cockroaches in the list of Carboniferous insects attached to Woodward's paper with a similar title.

——— See also **Borre, A. P. de.**

Jordan, Hermann und Meyer, C. E. H. von. Ueber die crustaceen der steinkohlenformation von Saarbrücken. (Palaeontogr., 4 : 1-15, pl. 1-2.) 4°. Cassel. 1854.

Describes, among other things, Adelophthalmus (Edrypterus) granosus, pp. 8-12, pl. 3, figs. 1-2, afterwards considered a cockroach by Goldenberg and others.

Kirkby, James W. On the remains of insects from the coal measures of Durham. (Geol. mag., 4 : 388-390, pl. 17, figs. 6-8.) 8°. London. 1867.

Describes and figures without names two or three orthopterous insects from near Claxheugh.

Kusta, J. Ueber eine Blattina aus der lubnaer gaskohle. 8°. Prag. 1883. 8 pp., 1 pl. (Sitzungsb. k. böhm. gesellsch. wiss., 1882 : 430-437, pl.) 8°. Prag. 1882.

Describes Blattina (Anthracoblattina) lubnensis, and makes extended comparisons.

——— Ueber einige neue, böhmische blattinen. 8°. Prag. 1883. 4 pp., figs. (Sitzungsb. k. böhm. gesellsch. wiss., 1883 : 211-215, figs.) 8°. Prag. 1883.

Three species are described and two named.

——— O blattině z lupkového uhlí z Třemošné u Plzně. (Věstn. král. české společ. nauk, 1888 : 387-393, fig.) 8°. Prag. 1888.

Describes Oryctoblattina arndti.

Lesquereux, Leo. Botanical and palaeontological report on the geological state survey of Arkansas. (Owen, Second rep. geol. reconn. Arkansas, pp. 295-399, pl. 1-6.) 8°. Philadelphia. 1860.

Contains description, p. 314 and figure, pl. 5, fig. 11, of Blattina venusta from Carboniferous rocks of Frog Bayou.

McIntire, E. S. See **Elrod, M. N., and McIntire, E. S.**

McLachlan, R. Note sur l'insecte fossile décrit par M. P. de Borre sous le nom de Breyeria borinensis. (Comptes rendus soc. ent. Belg., (2), xli : 5-6.) 8°. Bruxelles. 1877. (Ann. soc. ent. Belg., 20 : 36-37.) 8°. Bruxelles. 1877.

Considers the insect an ephemerid.

——— Did flowers exist during the Carboniferous epoch? (Nature, 19 : 554.) 4°. London. 1879.

Breyeria is an ephemerid.

McLachlan, R.—Continued.

—— Did flowers exist during the Carboniferous epoch? (Nature, 20: 5–6.) 4°. London. 1879.

Response to Mr. Wallace, disputing the lepidopterous nature of Breyeria borinensis.

Mahr, Carl Hermann. Beitrag zur kenntniss fossilen insecten der steinkohlen formation Thuringens. (Neues jahrb. f. mineral., 1870: 282–285, figs.) 8°. Stuttgart. 1870.

Description and figure of two species of Blattina from Ilmenau.

Meyer, C. E. H. von. See Jordan, H., und Meyer, C. E. H.

Mylacris packardii. (Rand. notes nat. hist., 2: 64.) 8°. Providence. 1885

Mere mention of the name of a Carboniferous cockroach from Rhode Island.

Rost, Woldemar. De filicum ectypis obviis in lithanthracum vettinensium lobeiunensiumque fodinis. Halæ. 16°. 1839. pp. 4, 32.

Germar's Blattina didyma is described as a fern under the generic name Dictyopteris.

Saporta, Marquis Gaston de. See Borre, A. P. de; also Section VII.

Scudder, S. H. Two new fossil cockroaches from the Carboniferous of Cape Breton. (Can. nat., (n. s.), 7: 271–272, figs. 1–2.) 8°. Montreal. 1874.

Describes Blattina beeri and B. bretonensis.

—— Note on the wing of a cockroach from the Carboniferous formation of Pittston, Penn. (Proc. Bost. soc. nat. hist., 19: 238–239.) 8°. Boston. 1878. (Scudder, Entom notes, 6: 35–36.) 8°. Boston. 1878.

Describes Blattina fascigera.

—— [Lettre à M. de Selys-Longchamps.] (Compt. rend. soc. ent. Belg., (2), xxi: 2.) 8°. Bruxelles. 1876. (Ann. soc. ent. Belg., 19, compt. rend., 2.) 8°. Bruxelles. 1876.

Expressing an opinion against the lepidopterous character of Breyeria. Translation by Selys.

—— Palæozoic cockroaches; a complete revision of the species of both worlds, with an essay toward their classification. (Mem. Bost. soc. nat. hist., 3: 23–134, pl. 2–6.) 4°. Boston. 1879.

Forms vol. 1, pp. 43–153, pl. 2–6, of the Fossil insects of North America.
The first attempt to classify any group of paleozoic insects of both worlds by characters drawn from the venation of the wing. More than

Scudder, S. H.—Continued.

sixty species (thirteen of them new) are described and figured; they are divided into two tribes and eleven genera, and separated as a whole from modern cockroaches under the name Palæoblattariæ. See also Geinitz, F. E.

—— [Exhibition of a Carboniferous cockroach.] (Psyche, 3: 186.) 4°. Cambridge. 1881.

Showing difference in venation of opposite wings.

—— A new and unusually perfect Carboniferous cockroach from Mazon Creek, Illinois. (Proc. Bost. soc. nat. hist., 21: 391–396.) 8°. Boston. 1882.

Describes Etoblattina mazona, the neuration of the wings of the two sides differing. Also separately printed, with a title-page and same pagination.

—— A gigantic walking-stick from the coal. (Science, 1: 95–96, fig.) 8°. Cambridge. 1883.

Notice of Titanophasma fayoli.

—— The species of Mylacris, a Carboniferous genus of cockroaches. (Mem. Bost. soc. nat. hist., 3: 299–309, pl. 27, figs. 5–11.) 4°. Boston. 1884.

Forms vol. 1, pp. 263–273, pl. 13, of the Fossil insects of North America.
Ten species are enumerated and tabulated and six described at length.

—— Dictyoneura and the allied insects of the Carboniferous epoch. (Proc. Amer. acad. arts sc., 20: 167–173.) 8°. Boston. 1885.

Systematic revision of the species which had been referred to Dictyoneura with descriptions of many new genera and species, separated by analytical tables.

—— Cockroaches from the Carboniferous period. (Proc. Bost. soc. nat. hist., 23: 356–357.) 8°. Boston. 1887.

Criticism of Woodward's conclusions concerning some British species.

—— An interesting paleozoic cockroach fauna, at Richmond, Ohio. (Proc. Bost. soc. nat. hist., 24: 45–53.) 8°. Boston. 1889.

Description of eight new species of Etoblattina.

—— New types of cockroaches from the Carboniferous deposits of the United States. (Mem. Bost. soc. nat. hist., 4: 401–415, pl. 31–32.) 4°. Boston. 1890.

Forms pp. 377–391, pl. 23–24, of the Fossil insects of North America, vol. 1. Ten species and two genera are described.

Scudder, S. H. See also Borre, A. P. de; Fontaine, W. M., and White, I. C.

Selys-Longchamps, M. E. de. See Scudder, S. H.

Smith, Sidney Irving. Notice of a fossil insect from the Carboniferous formation of Indiana. (Brief contributions to zoology from the museum of Yale College, no. ix.) 8°. [New Haven. 1871.] pp. 3, fig. (Amer. journ. sc., (3), 1: 44–46, fig.) 8°. New Haven. 1871.

Describes Paolia vetusta.

Wallace, A. R. Did flowers exist during the Carboniferous epoch? (Nature, 19: 582.) 4°. London. 1879.

Maintains the lepidopterous nature of Breyeria. See also McLachlan, R.

Weyers, J. L. See Borre, A. P. de.

White, I. C. See Fontaine, W. M., and White, I. C.

Woodward, H. Some now British Carboniferous cockroaches. (Geol. mag., n. s., dec. 3, vol. 4: 49–58, pl. 2.) 8°. London. 1887.

After a historical résumé of Carboniferous cockroaches, with figures of a few species, borrowed from Miall's work, three British species are described referred to Etoblattina, Lithomylacris, and Leptoblattina. See Scudder, S. H.

———. On the discovery of the larval stage of a cockroach, Etoblattina peachii H. Woodw. from the coal measures of Kilmaurs, Ayrshire. (Geol. mag., (3), 4: 433–435, pl. 12.) 8°. London. 1887.

Separate under the title: On Etoblattina peachii. Full description; placed provisionally in Etoblattina.

IIIe.—Hemipteroidea.

. See also under Section I and Section II.

Brauer, F. Betrachtungen über die verwandlung der insekten im sinne der descendenztheorie. 8°. Wien. 1869. pp. 21, pl. (Verhandl. k. k. zool.-bot. gesellsch. Wien., 19: 299–319, pl. 10.) 8°. Wien. 1869.

Contains, p. 19 [317], a slight reference to the structure of Eugereon.

Dohrn, A. Eugereon boeckingi, eine neue insectenform aus dem todtliegenden. 4°. Cassel. 1866. t. p., pp. 3, taf. (Palaeontogr., 13: 333–340, taf. 41.) 8°. Cassel. 1866.

Dohrn, A. - Continued.

Description and discussion of the affinities of the most remarkable fossil insect yet discovered, considered here to unite the Hemiptera and Neuroptera. Dictyoptera is proposed as an ordinal term to include it.

Geinitz, F. E. Versteinerungen aus dem brandschiefer der unteren dyas von Weissig bei Pillnitz in Sachsen. 8°. Stuttgart. 1873. pp. 14, pl. (Neues jahrb. f. miner., 1873: 691–704, taf. 3.) 8°. Stuttgart. 1873.

Describes a Fulgorina. See same title in Section IIId.

Suellen van Vollenhoven, Samuel Constant. Eugereon boeckingi. (Verslag alg. vergad. nederl. entom. vereen., 22: 13.) 8°. 's Gravenhage. 1867. (Tijdschr. v. entom., (2), 3: 13.) 8°. 's Gravenhage. 1868.

Denies that there are any purely hemipterous characteristics in Eugereon; but regards it as simply neuropterous.

IIIf.—Coleopteroidea.

. See also under Section I and Section II.

Brongniart, C. J. E. Note sur des perforations observées dans deux morceaux de bois fossile. (Ann. soc. ent. France, (5), 7: 215–220, pl. 7, ii.) 8°. Paris. 1877.

Describes the borings of a xylophagid allied to Hylesinus. Noticed by Dr. Hector George in the Feuilleton of Le constitutionnel for 21 Nov., 1877. See also Girard, M. See same title in Section Vf.

Dathe, E. Ueber schlesische eulmpetrefracten. (Zeitschr. deutsch. geol. gesellsch., 1885: 542–543.) 8°. Berlin. 1885.

Announces the discovery of beetle remains in the culm near Steinkunzendorf.

Geinitz, H. B. Die versteinerungen der steinkohlenformation in Sachsen. f°. Leipzig. 1855. pp. 7, 61, pl. 35.

Insecta, pp. 1–2, pl. 8, figs. 1, 4, are represented only by by borings of supposed coleoptera.

George, Hector. See Brongniart, C. J. E.

G[irard], M. Les perforations des bois fossiles. (La nature, 6: 112, figs. 1–6.) 4°. Paris. 1878.

Popular account of Brongniart's two papers on the subject with figures. See same title in Section Vf.

Rouchy, l'abbé. Découvertes de perforations de larves fossiles. (Petites nouv. entom., 1: 551.) 4°. Paris. 1875.

Borings of Coleoptera in a trunk of fossil Walchia.

IV.—GENERAL FOR MESOZOIC TIME.

* . * See also under Section I.

Assmann, A. Ueber die von Germar beschriebenen und im paläontologischen museum zu München befindlichen insekten aus dem lithographischen schiefer in Bayern. (Amtl. ber. versamml. deutsch. naturf., 50: 191–192.) 4°. München. 1877.

A brief statement of his views of the modern groups in which Germar's species should be placed.

Beckles, S. H. On the lowest strata of the cliffs at Hastings. (Quart. journ. geol. soc. Lond., 12, proc., 288–292; with a section.) 8°. London. 1856.

Merely mentions (p. 291) the discovery of insects in the upper members of the series, referred to the wealden.

Binfield, Henry. See Binfield, W. R. and H.

Binfield, William R. and Henry. On the occurrence of fossil insects in the wealden strata of the Sussex coast. (Quart. journ. geol. soc. Lond., 10, proc., 171–176.) 8°. London. 1854.

The insects are mentioned only by suborders, and the paper is mostly made up of geological sections of the places where insects were found.

Blake, J. F. See Tate, R., and Blake, J. F.

Bouvé, Thomas Tracy. See Deane, J.

Brauer, F., Redtenbacher, Jos., und Ganglbauer, Ludwig. Fossile insekten aus der juraformation Ost-Sibiriens. 4°. St. Petersburg. 1889. pp. 2, 22, pl. 2. (Mém. acad. sc. St.-Pétersb., (7), 36, no. 15.) 4°. St. Petersburg. 1889.

The authors describe and figure about a couple of dozen insects, mostly aquatic and larval forms. They regard Palæontina and Phragmatœcites as Cicadidæ.

Brodie, P. B. A notice on the discovery of the remains of insects, and a new genus of isopodous Crustacea belonging to the family Cymothoidæ in the wealden formation in the Vale of Wardour, Wilts. (Proc. geol. soc. Lond., 3: 134–135.) 8°. London. 1839. (Lond. Edinb. phil. mag., (3), 15: 534–536.) 8°. London. 1839.

A section of the locality with a general account of its fossil remains, those of the insects belonging to several orders. An extended notice will be found in the Neues jahrb. f. mineral., 1843: 258–239. 8°. Stuttgart. 1843.

Brodie, P. B.—Continued.

—— On the discovery of insects in the lower beds of lias of Gloucestershire. (Rep. brit. assoc. adv. sc., 1842, notices, 58.) 8°. London. 1843.

General notice of the finding of a few insect remains, mostly coleopterous, near Cheltenham.

—— Notice on the discovery of insects in the wealden of the Vale of Aylesbury, Bucks, with some observations on the distribution of these and other fossils in the Vale of Wardour, Wiltshire. (Ann. mag. nat. hist., 11: 480–482.) 8°. London. 1843.

Sufficiently described by the title.

—— Notice on the discovery of insects in the wealden in the Vale of Aylesbury, Bucks, with some additional observations on the wider distribution of these and other fossils in the Vale of Wardour, Wiltshire. (Lond. Edinb. Dubl. phil. mag., (3), 23: 512–514.) 8°. London. 1843. (Proc. geol. soc. Lond., 4: 780–782.) 8°. London. 1843.

Same as the preceding.

—— Notice on the discovery of the remains of insects in the lias of Gloucestershire, with some remarks on the lower members of this formation. (Lond. Edinb. Dubl. phil. mag., (3), 23: 529–531.) 8°. London. 1843. (Ann. mag. nat. hist., 11: 509–511.) 8°. London. 1843. (Proc. geol. soc. Lond., 4: 14–16.) 8°. London. 1842–43. (Athenæum, 1843: 40–41.) 4°. London. 1843.

Notice of the discovery and geological position of insects near Gloucester and Cheltenham, at Wainlode Cliff and at Westbury; an extension of the paper mentioned from the British association report.

ABSTRACT: Sur des débris d'insectes du lias du Gloucestershire. (L'institut, 1843, i, 47.) 4°. Paris. 1843.

ABSTRACT: Notiz über die entdeckung von insectenresten im lias von Gloucestershire mit einigen bemerkungen über die untern glieder dieser formation. (Neues jahrb. f. mineral., 1844: 127–129.) 8°. Stuttgart. 1844.

—— A history of the fossil insects in the secondary rocks of England.

Brodie, P. B.—Continued.

Accompanied by a particular account of the strata in which they occur, and of the circumstances connected with their preservation. 8°. London. 1845. pp. (18), 130, pl. 11.

The introductory observations, explanation of plates, notes, and many names, are by Westwood. This, the only separate work on fossil insects ever published in England, is still the chief source of our too inexact knowledge of the liassic and other secondary insects of that country. The body of the work, Mr. Brodie's part, is divided into four chapters, of which the first deals with the wealden, the second with the oolite, the third with the lias, and the fourth with miscellaneous matter, including insects of continental strata.

—— On the insect-limestone and its associate beds. (Murchison, Outl. geol. Cheltenham. 2d ed., 51–53.) 8°. London. 1845.

Simply a discussion of the mode of deposition of these rocks.

—— Notice of the existence of purbeck strata with remains of insects and other fossils, at Swindon, Wilts. (Quart. journ. geol. soc. Lond., 3, proc., 53–54.) 8°. London. 1847.

A geological paper, giving no further account of the insects than appears in the title.

—— On the insect beds in the purbeck formation of Dorset and Wilts; and a notice of the occurrence of a neuropterous insect in the Stonesfield slate of Gloucestershire. (Quart. journ. geol. soc. Lond., 9, proc., 344.) 8°. London. 1853.

Published only by title; probably same as next.

—— On the insect beds of the purbeck formation in Wiltshire and Dorsetshire. (Quart. journ. geol. soc. Lond., 10, proc., 475–492.) 8°. London. 1854.

Mostly occupied with geological sections, but p. 481 gives an account, in general terms, of the condition and character of the insects discovered, most of which were Coleoptera.

—— A sketch of the lias generally in England, and of the insect and saurian beds. (Proc. Warw. nat. arch. field club, 1868, pp. 1–24.) 8°. Warwick. 1868.

Mostly occupied with the geology of the insect-beds, but with occasional reference (especially on pp. 18–19) to the insects contained in them.

—— Notes on a railway-section of the lower lias and rhaetics between Stratford-on-Avon and Fenny Compton, on the occurrence of the rhaetics near Kineton,

Brodie, P. B.,—Continued.

and the insect beds near Knowle, in Warwickshire, and on the recent discovery of the rhaetics near Leicester. (Quart. journ. geol. soc. Lond., 30, 1: pp. 746–749.) 8°. London. 1874.

Simply notices the discovery of certain insects at Copt Heath near Knowle.

—— The lower lias at Eatington and Kineton, and on the rhaetics in that neighbourhood, and their further extension in Leicestershire, Nottinghamshire, Lincolnshire, Yorkshire, and Cumberland; . . . being a paper read at the annual meeting of the Warwickshire natural history and archaeological society, held at the museum, Warwick, April 2nd, 1875. 8°. Warwick. [n. d.] pp. 14. (Ann. rep. Warw. nat. hist. arch. soc.)

Principally occupied with geology, but with a few special references to insects, particularly on p. 10; separate only seen.

—— See also **Strickland, H. E.**

Buckman, James. On the occurrence of the remains of insects in the upper lias of the county of Gloucester. (Proc. geol. soc. Lond., 4: 211–212.) 8°. London. 1843. (Lond. Edinb. Dubl. phil. mag., 21: 377.) 8°. London. 1844.

Notices Æschna brodiei, without description, and the occurrence of two Coleoptera and a Tipula at Dumbleton.

Chambers, Victor Tousey. Burrowing larvæ. (Nature, 25: 629.) 4°. London. 1882.

Compares recent and fossil "mines" of lepidopterous larvæ, referring to Hagen's statement on a previous page.

Dana, J. D. See **Deane, J.**

Deane, James. On the sandstone fossils of Connecticut River. (Journ. acad. nat. sc. Philad., (2), 3: 173–178, pl. 18–20.) 4°. Philadelphia. 1856.

On pl. 19 are figured tracks of what the author presumes are articulated animals, in which he is supported by the opinions, quoted on p. 177, of Professors Leidy, Wyman, and Dana the latter believing them probably crustacean. Some are possibly the tracks of insects.

—— Ichnographs from the sandstone of Connecticut River. 4°. Boston. 1861. pp. 61, pl. 46.

Contains introduction, pp. 3–4, by A. A. Gould; biographical notice (of Dr. Deane) by H. I. Bowditch, pp. 5–14; a memoir upon the fossil footmarks and other impressions of the Connecticut River sandstone, by James Deane (compiled by

Deane, J.—Continued.

Thomas Tracey Bouvé, with a note by the compiler, p. 17, and the memoir, pp. 19-32; description of the plates, pp. 33-61 (by T. T. Bouvé).

References to insect tracks are made on p. 26, and in the descriptions of pl. 40-42 (pp. 57-58).

Deichmüller, J. V. Die insecten aus dem lithographischen schiefer im Dresdener Museum. (Mitth. k. min.-geol. praeh. mus. Dresd., heft 7, pp. 14 (8-14 marked 4-10), 88, pl. 1-5.) 4°. Cassel. 1886.

Describes at length over fifty species, of which thirty-six are figured with many rectifications of earlier authors.

Dupont, Édouard. Sur la découverte d'ossements d'iguanodon, de poissons et de végétaux dans la fosse Sainte-Barbe, du charbonnage de Bernissart. (Bull. acad. sc. Belg., (2), 46: 387-408.) 8°. Bruxelles. 1878.

Mentions, p. 395, the occurrence of insect larvæ in this wealden deposit.

Emmons, Ebenezer. American geology, containing a statement of the principles of the science, with full illustrations of the characteristic American fossils, with an atlas and a geological map of the United States. Part vi. 8°. Albany. 1857. pp. 10, 152, pl. [13].

P. 136 compares trails found in Trias, some of which he figures, with those of living insects.

Fritsch, A. O hmyzech v českém útvaru křídovém. (Vesmír, 13: 205-206, figs.) 4°. v Praze. 1884.

Briefly notices and figures elytra of four beetles, eggs of a sawfly, and larval cases of a caddis fly and a tineid from the cretaceous of Bohemia.

Reference is made to an earlier notice of the beetles named Silphites in the Archiv pro vyzkum čech., díl. 1. odd. 2, str. 170, which I have not seen.

Fromont [Louis]. [Empreintes sur une plaque de pierre lithographique.] (Ann. soc. ent. Belg., 23, comptes rendus, 35.) 8°. Bruxelles. 1880.

Mention of impressions considered to resemble the antennæ of an insect.

Ganglbauer, Ludwig. See Brauer, F., Redtenbacher, J., und Ganglbauer, L.

Geinitz, F. E. Der jura von Dobbertin in Mecklenburg und seine versteinerungen. (Zeitschr. deutsch. geol. gesellsch., 32: 510-535, taf. 22.) 8°. Berlin. 1880.

Geinitz, F. E.—Continued.

Contains, pp. 519-531, Insectenfauna des dobbertiner jurā, in which seventeen insects are described, the greater part of them new. The plate is wholly devoted to insects.

—— Die flötzformation Mecklenburgs. 8°. Güstrow. 1883. 150 pp., map, 6 pl. (Arch. ver. freunde naturg. Mecklenb., 37: 1-151.) 8°. Güstrow. 1883.

A few insects are described on pp. 29-31 and figured on pl. 6.

—— Ueber die fauna des dobbertiner lias. (Zeitschr. deutsch. geol. gesellsch., 1884: 566-583, pl. 13.) 8°. Berlin. 1884.

Describes or mentions over forty species, the Coleoptera only by reference to similar forms from the Swiss and English Lias. Twenty-seven specimens are figured.

Germar, E. F. Ueber die versteinerten insecten des juraschiefers von Solenhofen aus der sammlung des grafen zu Münster. (Oken, Isis, 1837: 421-424.) 4°. Leipzig. 1837.

Germar compares the few insects then known from Solenhofen with the tertiary insects, and concludes that: 1°, none of the jurassic species are identical with the living; 2°, there are no strikingly strange forms; 3°, the general facies of the fauna is that of middle Europe and the United States, and indicates a similar climate; 4°, all are wood or leaf eaters, excepting some water beetles and a Geotrupes. This paper appears to have been read before the Jena meeting of the Deutscher naturforscher und ärtzte, in 1836, but I have been unable to consult the report of that meeting.

—— Die versteinerten insecten Solenhofens. (Nova acta acad. leop.-carol., 19, i: 187-222, tab. 21-23.) 4°. Vratislaviae et Bonnae, 1839.

Describes and rudely figures seventeen insects of various orders, of which eleven are credited to Münster. The descriptions are preceded by some general remarks, historical and otherwise, upon Solenhofen and other fossil insects.

—— Beschreibung einiger neuen fossile insecten (i.) in den lithographischen schiefern von Bayern und (ii.) in schieferthon des steinkohlengebirges von Wettin. (Münst., Beitr. z. petref., 5: 79-94, taf. 9, 13.) 4°. Bayreuth. 1842.

The first part, pp. 79-90, pl. 9, 13, describes and figures nine insects of various orders from Solenhofen, being the first memoir on the subject. See the same title in Section II.

—— See also **Assmann, A.**

Goss, H. Three papers on fossil insects, and the British and foreign formation in which insect remains have been detected. No. 2. The insect fauna of the secondary or mesozoic period. 8°. [London. 1879.] pp. 37. (Proc. geol. assoc., 6, iii: 116-150.) 8°. London. 1879.

ABSTRACT: The insect-fauna of the secondary or mesozoic period, and the British and foreign strata in which insect remains have been detected. (Geol. mag., (n. s.), 5: 134-136.) 8°. London. 1878.

See same general title in Section II and Section VI, with note in former.

—— Introductory papers on fossil entomology. No. 6. Mesozoic time. On the insects of the triassic period, and the animals and plants with which they were correlated. (Entom. monthl. mag., 15: 245-246.) 8°. London. 1879.

The same. No. 7, part 1. Mesozoic time. On the insects of the jurassic period, and the animals and plants with which they were correlated. (Entom. monthl. mag., 16: 7-10.) 8°. London. 1879.

The same. No. 7, part 2. Mesozoic time. On the insects of the jurassic period and the animals and plants with which they were correlated. (Entom. monthl. mag., 16: 25-29.) 8°. London. 1879.

The same. No. 8. Mesozoic time. On the insects of the cretaceous period and the animals and plants with which they were correlated. (Entom. monthl. mag., 16: 58-60.) 8°. London. 1879.

See same title in Section I, Section II, and Section VI.

Gray, O. W. See **Walling,** H. F., and **Gray,** O. W.

Hagen, H. A. Fossil insects of the Dacota group. (Nature, 25: 265-266.) 4°. London. 1882.

A brief note recording the discovery of galls and mines in fossil leaves from Kansas and Nebraska.

Hasselt, A. W. M. van. See **Weyenbergh,** H.

Heer, O. 1. [Zwei] geologische vorträge gehalten im März 1852 von Oswald Heer und A. Escher von der Linth. 1. Die lias-insel des Aargau's. [*Entitled on cover:* Ueber die lias-insel im Aargau.] 2. Ueber die gegend von Zürich in der letzten periode der vorwelt, mit einer blockkarte der Schweiz. 4°. Zürich. [1852.] pp. 28, pl. 2.

Heer, O.—Continued.

Heer's portion, pp. 1-15, pl. 1, is largely devoted to insects, the greater part of which are woodboring Coleoptera, and indicate a warm tropical climate. Twenty-two species are described and figured, of which nineteen are beetles.

—— Die kreide-flora der arctischen zone, gegründet auf die von den schwedischen expeditionen von 1870 und 1872 in Grönland und Spitzbergen gesammelten pflanzen. 4°. Stockholm. 1874, pp. 138, pl. 38. (Kongl. svenska vetensk. akad. handl., 12, vi.)

Insekten der kreideschichten, pp. 91-92, pl. 17, describes two Coleoptera. Myriopoden, pp. 120-121, pl. 33, describes Julopsis cretacea.

Forms vol. 3, no. 2, of Heer's Flora fossilis arctica.

—— Flora fossilis helvetica. Die vorweltliche flora der Schweiz. 4°. Zürich. [1876-] 1877. t. p., pp. 6, 182, pl. 70.

Describes, p. 76, and figures, pl. 27, a single beetle from the kemper of Rüdhard, canton Basel, and a neuropteron on p. 77, pl. 29, from the trias of Mythen, canton Schwyz.

Hislop, S. On the age of the fossiliferous thin-bedded sandstone and coal of the province of Nágpur, India. (Quart. journ. geol. soc. Lond., 17, i: 346-354.) 8°. London. 1861.

Refers to the discovery of Blattariæ and Coleoptera at Kotá, probably liassic.

—— Supplemental note on the plant-bearing sandstones of central India. (Quart. journ. geol. soc. Lond., 18, i: 36.) 8°. London. 1862.

Discovery of more insects at Kotá.

Hitchcock, Charles Henry. See **Hitchcock,** E.; **Walling,** H. F., and **Gray,** O. W.

Hitchcock, Edward. Ichnology of New England. A report on the sandstone of the Connecticut valley, especially its fossil footmarks, made to the government of the commonwealth of Massachusetts. 4°. Boston. 1858. pp. 12, 220, pl. 60.

Refers to prints, supposed to be those of insects, on pp. 147-160, 165-166, 188-189, and mentions an insect larva, pp. 7-3. The figures of these are distributed on plates 24, 27-31, 42.

—— Supplement to the Ichnology of New England, a report to the government of Massachusetts in 1863. 4°. Boston. 1865. pp. 10, 96, pl. 20.

Appendix B. Descriptive catalogue of the specimens in the Hitchcock ichnological cabinet of

Hitchcock, E.—Continued.

Amherst college, prepared by C. H. Hitchcock. pp. 41–88. Tracks of insects, pp. 13–17; tracks of myriapods, pp. 17–18.

Leidy, Joseph. See **Deane, J.**

Moore, Charles. On the palæontology of the middle and upper lias. (Proc. Somersetsh. archæol. nat. hist. soc., 2: 61–76.) 8°. Taunton. 1865–1866.

His collection of lias insects consists of about 1,000 specimens. "The families represented at Ilminster include Libellula, Neuroptera, Orthoptera, Homoptera, Diptera, and Coleoptera."

Morris, J. On some sections in the oolitic district of Lincolnshire. (Quart. journ. geol. soc. Lond., 9, i: 317–344, pl. 14.) 8°. London. 1853.

Mentions the occurrence of insects on p. 324 and note.

Münster, G. Nachtrag zu dem aufsätze des professor Germar in theil 4, heft 2, dieser zeitschrift über die versteinerungen von Solnhofen. (Teutschl. geogn. geol. dargest., 5: 578–581.) 16°. Weimar. 1829.

Gives a list of the fossils known to him, among which, on p. 579, occurs "19 arten insecten darunter 2 arten libellen." Germar's paper referred to has nothing on insects.

——— Insekten in lias. (Neues jahrb. mineral., 1835: 333.) 8°. Stuttgart. 1835.

Has discovered lias insects in the neighborhood of Brzezina.

——— See also **Germar, E. F.**

Oppenheim, P. Die insectenwelt des lithographischen schiefers in Bayern. (Palæontogr., 34: 215–247, pl. 30–31.) 4°. Stuttgart. 1888.

Description of about forty-five species, including several new genera; nearly half of them are Coleoptera and the remainder distributed among the other orders excepting Lepidoptera and Diptera. A full discussion of the Rhipidorhabdi and their systematic position closes the paper.

· Phillips, John. The neighborhood of Oxford, and its geology. (Oxford essays, 1855: 192–212.) 8°. London. 1855.

A paragraph, p. 204, is given to the insects of the Stonesfield slate.

——— Geology of Oxford and the valley of the Thames. 8°. Oxford. 1871. pp. 24, 523, pl. 17.

Contains, p. 123, fossils of the liassic period, pp. 173–174, fossils of the Stonesfield beds, in both of which insects are referred to.

Redtenbacher, Jos. See **Brauer, F.**

Redtenbacher, J., and Ganglbauer, L.

Scudder, S. H. The insects of the triassic beds at Fairplay, Colo. (Mem. Bost. soc. nat. hist., 4: 457–472, pl. 41–42. 4°. Boston. 1890.

Forms pp. 433–448, pl. 33–34, of the Fossil insects of North America, Vol. 1. The fauna consists almost exclusively of cockroaches, of which eighteen species of seven genera are described and figured, besides three Hemiptera.

Strickland, Hugh Edwin. On the results of recent researches into the fossil insects of the secondary formations of Britain. (Rep. Brit. assoc. adv. sc., 1845, notices, 58.) 8°. London. 1846.

A general account of what had been published by Brodie, with a few general deductions.

Tate, Ralph, and Blake, J. F. The Yorkshire lias. 8°. London. 1876. pp. 12, 475, 12, pl. 19, 4, map.

Class Insecta by J. F. Blake, p. 426, pl. 16 (pars). Figures without description a Buprestites and a Chauliodites.

Walling, Henry F., and Gray, O. W. Official topographical atlas of Massachusetts from astronomical, trigonometrical, and various local surveys, compiled and corrected by H. F. Walling and O. W. Gray. Folio. [Boston?] 1871. 100 pp.

The geological description on pp. 17–23 by C. H. Hitchcock, contains, p. 21, a list of Ichnozoa.

Warren, John Collins. Remarks on some fossil impressions in the sandstone of Connecticut River. 8°. Boston. 1854. pp. 54, pl.

On p. 37 he refers some of the impressions as perhaps made "by the feet and bodies of large insects."

Westwood, J. O. Description of the remains of fossil insects from the Purbeck formation of Dorset and Wilts, and from the Stonesfield slate of Gloucestershire. (Quart. journ. geol. soc. Lond., 9, proc., 344.) 8°. London. 1853.

Unpublished; apparently the same as the next.

——— Contributions to fossil entomology. (Quart. journ. geol. soc. Lond., 10: 378–396, pl. 14–18.) 8°. London. 1854.

About one hundred and fifty specimens are figured and fifty-nine species named. They are nearly all from Purbeck strata, about half of them Coleoptera, and the remainder are referred mostly and about equally to Hemiptera, Orthoptera and Neuroptera. The separata have a title on reverse of p. 378.

——— See also **Brodie, P. B.**

Weyenbergh, H. Sur les insectes fossiles du calcaire lithographique de la Bavière, qui se trouvent au Musée Teyler. 8°. Harlem. 1869. t. p., pp. 48, pl. 4. (Arch. mus. Teyl., 2: 247-294, pl. 34-38.) 8°. Harlem. 1869.

Describes forty-eight species, many of them new; preceded by a list of the sixty previously known jurassic hexapods, and followed by five pages of general considerations.

—— Prodromus en algemeene beschouwing der fossiele insekten van Beijeren. 8°. ['s Gravenhage.] 1869. pp. 19. (Tijdschr. entom., (2), 4: 230-248.) 8°. 's Gravenhage. 1869.

A list of one hundred and four insects is given, followed by general remarks, including, pp. 12-14 (241-243), a comparison of the secondary insects of England and Bavaria; pp. 231-234 are printed 131-134.

—— Notes sur quelques insectes du calcaire jurassique de la Bavière. 8°. Harlem. 1873. t. p., pp. 7. (Arch. mus. Teyl., 3: 234-240.) 8°. Harlem. 1873.

Further discussion of the affinities of four species included in the preceding paper, especially of Hassellides primigenius (with the opinions of van Hasselt on this species) and of Sphinx snelleni.

—— Varia zoologica et palaeontologica. (Períód. zool., org. soc. entom. argent., 1: 77-111, lam. 2-3.) 8°. Buenos Aires. 1874.

Weyenbergh, H.—Continued.

"Insectes fossiles," pp. 81-107, lam. 3, includes descriptions and discussion of a half-dozen Solenhofen insects, of which two or three are new, followed by the list mentioned under the next entry, and a list, p. 107, of the secondary insects of Bavaria not represented in the Musée Teyler.

—— Enumération systématique des espèces qui forment la faune entomologique de la période mésozoïque de la Bavière; en même temps Supplément du Catalogue de la collection paléontologique du Musée Teyler. 8°. [Buenos Aires, 1874.] pp. 20. (Períód. zool., org. soc. entom. argent., 1: 87-106.) 8°. Buenos Aires. 1874.

Contains two hundred and sixty-five numbers, of which about thirty are undetermined; full references to descriptions and illustrations are added.

—— Sur les chenilles fossiles. (Pot. nouv. entom., vol. 2, no. 202, p. 253.) 4°. Paris. 1878.

A note calling attention to the caterpillar of the jurassic Sphinx snelleni previously described by him.

Wyman, Jeffries. See Deane, J.

Yxem, E. Versteinerte insecten-zellen. (Ber. naturw. ver. Harzes, 1840-46, 2e auff., p. 26.) 4°. Wernigerode. 1856.

Exhibition of drawings of insect cells like beecomb from jurassic (?) beds at Chausseebau near Harzleben.

V.—SPECIAL FOR MESOZOIC TIME.

Va.—Mesozoic Myriapoda.

. See under Section I and Section IV, there being no separate titles under this head.

Vb.—Mesozoic Arachnida.

. See also under Section I and Section IV.

Kundmann, Johann Christian. Rariora naturæ et artis, item in re medica, oder Seltenheiten der natur und kunst des kundmannischen naturalien-cabinets, wie auch in der arzneywissenschaft. f°. Breslau und Leipzig. 1737. 2 t. p., fl. (8), col. 1312 (=ff. 328), ff. (17), portr., figs., pl. 17.

Contains, col. 229-230, tab. 12, figs. 13-14, art. 28: Von einem geglaubten und wahren spinnen-steine, in which supposed spiders from the jurassic rocks of Eichstädt are figured.

Meyer, C. E. H. von. Vogel-federn und Palpipes priscus von Solenhofen. (Neues jahrb. mineral., 1861: 561.) 8°. Stuttgart. 1861.

A brief notice of seven new specimens of Palpipes, then regarded as an arachnid.

—— Zu Palpipes priscus aus dem lithographischen schiefer in Bayern. (Palaeontograph., 10: 299-304, taf. 50, figs. 1-4.) 4°. Cassel. 1863.

A detailed description of this supposed arachnid from several specimens.

Roth, Johann Rudolph. Ueber fossile spinnen des lithographischen schiefers. (Gel. anz. bay. akad. wiss., 32: 164-167, fig.) 4°. München. 1851. (Neues jahrb. mineral., 1851: 375-377, pl. 4 B, fig. 8.) 8°. Stuttgart. 1851.

Roth, J. R.—Continued.

Describes two species of a new genus, Palpipes, regarded as an arachnid.

Seebach, K. von. Ueber fossile phyllosomen von Solenhofen. (Zeitschr. deutsch. geol. gesellsch., 23: 340–346, pl. 8.) 8°. Berlin. 1873.

Shows that Phalangites or Palpipes is a crustacean, and not, as formerly supposed, an arachnid.

Vc.—Mesozoic Neuroptera.

. See also under Section I and Section IV.

Brauer, F. Verzeichniss der bis jetzt bekannten neuropteren im sinne Linne's. pp. 90. 8°. [Wien.] [n. d.] (Verhandl. k. k. zool.-bot. gesellsch. Wien, 18: 359–416, 711–742.) 8°. Wien. 1868.

Includes the fossil genera and species, and contains, p. 738 (80), a list of the fossil Libellulina.

Brodie, P. B. Notice on the discovery of a dragon-fly and a new species of Leptolepis in the upper lias near Cheltenham, with a few remarks on that formation in Gloucestershire. 8°. pp. 4, pl. (Quart. journ. geol. soc. Lond., 5, proc., 31–37, pl. 2.) 8°. London. 1849.

The description (2 pp.) is by Westwood, but the name, Libellula (Heterophlebia) dislocata, is by Brodie. The rest of the paper is on the geology of the district. I have not seen the separate paper.

Buch, Christian Leopold von. See **Erichson, W. F.**

Buckland, W. On the discovery of a fossil wing of a neuropterous insect in the Stonesfield slate. (Proc. geol. soc. Lond., 2: 688.) 8°. London. 1835. (Lond. Edinb. Dubl. phil. mag., (3),13: 388.) 8°. London. 1838.

Brief notice of Hemerobioides giganteus.

ABSTRACT: Découverte d'une aile fossile d'insecte névroptère dans les schistes de Stonesfield. (Rev. zool., 1839: 29.) 8°. Paris. 1839.

Abstract by Malepeyre.

Buckman, J. Remarks on Libellula brodiei (Buckman), a fossil insect from the upper lias of Dumbleton, Gloucestershire. (Ann. mag. nat. hist., (2), 12: 436–438.) 8°. London. 1853.

Claims Æschna brodiei, Libellula (Heterophlebia) dislocata, and Agrion buckmani to be one insect which should bear the name in title.

Charpentier, Toussaint von. Libellulinae europaeae descriptae ac depictae cum tabulis 48 coloratis. 4°. Lipsiae. 1840. t. p., pp. 181, pl. 48.

Under the section De libellulinis petrefactis, pp. 170–173, pl. 48, the author gives a résumé of what was then known on the subject, and describes and figures some new forms. All mentioned are from the secondary rocks. In the explanation of the plates, p. 180, he speaks of fig. 1 as Libellulites solenhofiensis, which Kirby has wrongly quoted as a specific term, but which the context shows was merely meant to designate its source.

Dale, James Charles. Notes on some libellulae. (Ann. mag. nat. hist., 9: 257–258.) 8°. London. 1842.

Suggests that "Æshna lassalue" Strickland is nearer Cordulegaster or Petalura.

Dana, J. D. Fossil larva in the Connecticut River sandstone. (Amer. journ. sc. arts, (2), 33: 451–452.) 8°. New Haven. 1862.

Quotes an opinion from Dr. J. L. Leconte that Hitchcock's figure of Mormolucoides articulatus resembles the larva of an ephemerid; and the consequent wish of Dr. E. Hitchcock that the name should be changed to Palephemera medigæva.

Eichwald, Édouard d'. Sur un terrain jurassique à poissons et insectes d'eau douce de la Sibérie orientale. (Bull. soc. géol. France, (2), 21: 19–25.) 8°. Paris. 1864.

Describes Ephemeropsis trisetalis, pp. 21–22. The deposit was thought by Müller to be Eocene.

TRANSLATION: On a jurassic deposit containing fresh-water fish and insects in Eastern Siberia. (Quart. journ. geol. soc. Lond., 20, ii : 21–22.) 8°. London. 1864.

An abstract by R[alph] T[ate].

——— Ueber fossile insecten und belemniten. (Amtl. ber. vers. deutsch. naturf., 39: 169–172.) 4°. Giessen. 1865.

Notices, p. 170, the ephemerid larva described by him as Ephemeropsis, found in calcareous schists on the banks of the Turga in Siberia.

——— Lethaea rossica, ou Paléontologie de la Russie décrite et figurée. Text, 8°. Atlas, 4°. 3 v. Stuttgart. 1853–68. Vol. 1. Ancienne période (in 2 parts), pp. 1–19, 17–26, 1–681, 681–1657+titles to parts. 1860. Atlas. t. p., pp. 8, tab. 59. 1859.—Vol. 2. Période moyenne (in 2 parts), pp. 1–35, 1–640, 641–1304+titles to parts. 1865, 1868. Atlas. t. p., tab. 40, expl. of plates opp. plates. 1868.—Vol. 3.

Eichwald, E. d'.—Continued.

Période moderne, pp. 19,533. 1853. Atlas. t. p., pp. 4, tab. 14, expl. of plates opp. plates. 1853.

The few insects are contained in vol. 2, ii, pp. 1191-1195, tab. 37 (1868).

Erichson, Wilhelm Ferdinand. Zur abbildung der libelle von Solenhofen. (Buch, Jura in Deutschl., p. 135, pl. (3). Abhandl. kön. akad. wiss. Berlin, 1837, phys. abhandl.) 4°. Berlin. 1839.

Considers the insect figured by von Buch as partaking of the characters of the genera Æschna and Libellula. It was afterwards named Anax buchi by Hagen.

Fritsch, A. Palaeontologische untersuchungen der einzelnen schichten in der böhmischen kreideformation. (Archiv naturw. landesdurchf. Böhmen, bd. 1, sect. 2, pp. 181-256, pl. 3.) 8°. Prag. 1869.

Refers on p. 187 to the discovery of an elytron of a beetle, and a tube of a phryganid larva in clay schists at Kounic.

Germar, E. F. See **Hagen, H. A.**

Giebel, C. G. Zur fauna des lithographischen schiefers von Solenhofen. (Zeitschr. gesammt. naturw., 9 : 373-388, taf. 5-6.) 8°. Berlin. 1857.

Contains long descriptions and figures of two dragon-flies.

——— Eine neue Æschna aus den lithographischen schiefer von Solenhofen. (Zeitschr. gesammt. naturw., 16 : 127-131. taf. 1.) 8°. Berlin. 1860.

Describes very fully Æschna wittei.

Hagen, H. A. Ueber die fossile odonate Heterophlebia dislocata Westwood, nebst abbildung. (Stett. ent. zeit., 10 : 226-231. pl. 1.) 16°. Stettin. 1849.

An extended description, showing that it represents a new genus of Gomphidae. Dr. Hagen informs me that the most important vein is given in the wrong place by the lithographer.

——— A comparison of the fossil insects of England and Bavaria. (Entomol. annual, 1862, pp. 1-10.) 16°. London. 1862.

Devoted almost exclusively to a comparison of the Neuroptera of the Bavarian jura and the English lias-insects, by which he concludes the two faunas to be "extremely closely allied," and to be very different from the tertiary or existing forms.

Hagen, H. A.—Continued.

——— Comparison of fossil insects of England and Bavaria. (Report Brit. assoc. adv. sc., 31, notices, 113-114.) 8°. London. 1862.

Dealing mostly with Odonata. The same given more fully in the Entom. annual. See preceding entry.

——— Ueber die neuroptern aus dem lithographischen schiefer in Bayern. (Palaeontogr., 10 : 96-145, taf. 13-15.) 4°. Cassel. 1862.

An introduction of nine pages, containing besides other interesting matter the comparison of the mesozoic insects of England and Bavaria given the previous year in England (see the preceding entries), is followed by a list of thirty-seven species, mostly Odonata, found at Solenhofen and Eichstädt, by five pages of a review of earlier writers, especially Germar, and by the extended description of twenty-four species, pp. 114-145.

——— Notes on Tarsophlebia westwoodii Giebel, a fossil dragon-fly. (Entom. monthl. mag., 1 : 160-161.) 8°. London. 1864.

Heterophlebia believed to belong to the Gomphidae, Tarsophlebia to the Calopterygidae. The latter differs from all living Odonata in the length of the first tarsal joint.

——— Die Neuroptera des lithographischen schiefers in Bayern. Pars 1: Tarsophlebia, Isophlebia, Stenophlebia, Anax. 4°. Cassel. 1866, pp. 40, taf. 4. (Palaeontogr., 15 : 57-96, taf. 1-4.) 4°. Cassel. 1866.

Extended generic and specific descriptions of eight dragon-flies.

Herold, Johann Moritz David. See **Koehler, F.**

Hessel, Johann Friedrich Christian. See **Koehler, F.**

Hitchcock, E. See **Dana, J. D.**

Joly, Nicolas. Incontestablement, le Prosopistoma de Latreille est un éphémérien. (Mém. acad. sc. Toulouse, (8). 2 : 188-189.) 8°. Toulouse. 1880.

At the end of his paper, p. 189, he refers to this genus a secondary fossil figured by Brodie.

Kirby, W. F. See **Charpentier, T. von.**

Koehler, Friedrich. Ueber den libellulit von Solenhofen. (Zeitschr. f. mineral. [Taschenb. ges. mineral., jahrg. 20] bd. 2: 231-233, pl. 7, fig. 3.) 16°. Frankfurt a. M. 1826.

With note by Hessel giving the opinion of Herold. The insect is referred to Æschna.

Koehler, F.—Continued.

TRANSLATION: Account of a libellulite found at Solenhofen. (Edinb. new phil. journ., 2: 195, pl. 3, fig. 4.) 8°. Edinburgh. 1826.

The note is not appended.

LeConte, J. L. See also Dana, J. D.

Linden, Pierre Léonard van der. Notice sur une empreinte d'insecte, renfermée dans un échantillon de calcaire schisteux de Sollenhoven, en Bavière. 4°. (Bruxelles. 1827.) pp. 9, pl. (Nouv. mém. acad. roy. sc. Brux., 4: 245–253, pl.) 4°. Bruxelles. 1827.

Describes and figures Æschna antiqua.

Malepeyre. See Buckland, W.

Müller, Johannes. Fossile fische. (Middendorff's Sibirische reise, I. i : 259–263, pl. ii.) 4°. St. Petersburg. 1848.

Refers in three lines, p. 261, to the ephemerid larva found by Middendorff, which is figured pl. 11, fig. 7.

Packard, A. S. [On insect-remains occurring in nodules . . . north of Turner's Falls.] (Bull. Essex inst., 3: 1–2.) 8°. Salem. 1871.

Considers Palephemera medieva an aquatic coleopterous insect "belonging perhaps near the family Heteroceridae."

Phillips, J. Oxford fossils; No. 2. (Geol. mag., 3, 97–99, pl. 6.) 8°. London. 1866.

Describes and figures Libellula westwoodii from the Stonesfield slate, and compares it with Æschna brodiei from the lias.

Schmidel, Kasimir Christoph. Fortgesezte vorstellung einiger merkwürdigen versteinerungen mit kurzen anmerkungen versehen. 4°. Nürnberg. 1782. t. p., pp. 45, pl. 8–24.

Describes, p. 36, and figures, pl. 19, fig. 2, a dragonfly from Solenhofen, which he compares with Libella grandis.

Scudder, S. H. On Mormolucoides articulatus. (Proc. Bost. soc. nat. hist., 11: 140.) 8°. Boston. 1867.

Considers this triassic species to be a coleopterous larva.

——. The oldest known insect larva, Mormolucoides articulatus, from the Connecticut River rocks. (Mem. Bost. soc. nat. hist., 3: 431–438, pl. 45.) 4°. Boston. 1886.

Forms vol. i, pp. 323–330, pl. 19, of the Fossil Insects of North America.

Mormolucoides regarded as the larva of a sialid neuropteron.

Selys-Longchamps, M. E. de. Synopsis des Agrionines, 5me légion : Agrion (suite et fin). Les genres Telebasis, Argiocnemis et Hemiphlebia. 12°. Bruxelles. 1877. 65 pp. (Bull. acad. sc. Belg., (2), 43 : 97–159.) 12°. Bruxelles. 1877.

On p. 64 under the description of Hemiphlebia, he compares its structure with Tarsophlebia from the Jura.

——. Palæophlebia, nouvelle légion de caloptérygines. 8°. Benxelles, 1889. pp. 7. (Ann. soc. ent. Belg., 33, comptes rendus, 153–158, pl. 2.) 8°. Bruxelles. 1889.

Refers to it Heterophlebia, Stenophlebia, and a living Japanese form.

Strickland, H. E. On the occurrence of a fossil dragon-fly in the lias of Warwickshire. 8°. [London.] 1840. pp. 2, figs. 1–3. (Mag. nat. hist., (n. s.), 4 : 301–302, figs. 11–13.) 8°. London. 1840.

Description of "Æshna Hassina."

Tate, R. See Eichwald, E. d'.

Westwood, J. O. See Brodie, P. B.

White, Charles Abiathar. Report on the paleontological field work for the season of 1877. (Ann rep. U. S. geol. surv. terr., 11, 161–272.) 8°. Washington. 1879. [1880.]

Contains a notice of Corydalites fecundum and its geological position.

Woodward, H. On the wing of a neuropterous insect from the cretaceous limestone of Flinders River. North Queensland, Australia. (Geol. mag., (3), 1 : 337–339, pl. 11, fig. 1.) 8°. London. 1884.

Describes and figures Æschna flindersiensis.

Vd.—Mesozoic Orthoptera.

⁎⁎ See also under Section 1 and Section IV.

Heyden, Carl von. Chrysobothris veterana und Blabera avita, zwei fossile insekten von Solenhofen. (Palæontogr., 1 : 99–101, pl. 12, fig. 4–5.) 4°. Cassel. 1847.

Simple description. See same title in Section Vf.

Scudder, S. H. Triassic insects from the Rocky Mountains. (Amer. journ. sc., (3), 28 : 199–203.) 8°. New Haven. 1884.

General account of the cockroaches found near Fairplay, Col., and their relation to older and to modern types.

Scudder, S. H.—Continued.

—— Notes on mesozoic cockroaches. I. Pterinoblattina, a remarkable type of Palaeoblattariæ; II. Triassic Blattariæ from Colorado; III. On the genera hitherto proposed for mesozoic Blattariæ. (Proc. acad. nat. sc. Philad., 1885: 105–115.) 8°. Philadelphia. 1885.

The genera Pterinoblattina (with Gap.), Neorthroblattina (4 sp.), and Scutinoblattina (3 sp.), are described, and the old genera Blattidium, Elisama, Rithma, and Mesoblattina characterized.

—— A review of mesozoic cockroaches. (Mem. Bost. soc. nat. hist., 3: 439–485, pl. 46–48.) 4°. Boston. 1886.

Forms vol. 1, pp. 331–376, pl. 20–22, of the Fossil insects of North America.

Thirteen genera and seventy-nine species, many, especially British, species new.

Ve.—Mesozoic Hemiptera.

. See also under Section I and Section IV.

Buckton, G. B. Monograph of the British Aphides. 4 vol. 8°, London, 1875–1883.

See same title in Section I and Section VIIe.

Butler, Arthur Gardiner. On fossil butterflies, 4°. [London, 1873.] pp. 126–128, pl. 48. (Butl., Lepid. exot., part xv, pp. 126–128, pl. 48.) 4°. London. 1873.

Description and illustration of three species, including for the first time Palæontina oolitica. The figure of the latter was copied into the Graphic, of Feb. 22, 1873, with a brief account of it, under the title: "The oldest fossil butterfly in the world." A similar account appeared in Hardw. sc. gossip, 1873: 260–261, under the title: "The oldest fossil butterfly."

—— Notes on the impression of Palæontina oolitica in the Jermyn street museum. The cover entitled: On a fossil butterfly in the Museum of practical geology, Jermyn street. 8°. [London. 1874.] pp. 4, pl. (Geol. mag., dec. 2, vo. 11, pp. 446–449, pl. 19.) 8°. London. 1874.

A rejoinder to Scudder, and in favor of the lepidopterous character of Palæontina.

Scudder, S. H. [On an English fossil insect described as lepidopterous.] (Proc. Bost. soc. nat. hist., 16: 112,) 8°. Boston. 1874.

Considers Palæontina oolitica Butler as homopterous rather than lepidopterous. See also Butler, A. G.

Vf.—Mesozoic Coleoptera.

. See also under Section I and Section IV.

Brodie, P. B. Notice of the occurrence of an elytron of a coleopterous insect in the Kimmeridge clay at Ringstead Bay, Dorsetshire. (Quart. journ. geol. soc. Lond., 9, proc., 51–52.) 8°. London. 1853.

No further details of the insect than are given in the title.

Brongniart, C. J. E. Rapport sur un morceau de bois fossile trouvé dans le gault, terrain crétacé de Lottinghem (Pas-de-Calais). (Ann. soc. ent. France, (5), 6, bull. ent., 117–118.) 8°. Paris. 1876.

Refers the borer reported by Lartigne to Bostrychus. See also Lartigne.

—— Note sur des perforations observées dans deux morceaux de bois fossile. (Ann. soc. ent. France, (5), 7: 215–220, pl. 7, ii.) 8°. Paris. 1877.

Describes the borings of a xylophagid allied to Bostrychus. Noticed by Dr. Hector George in the Feuilleton of Le constitutionnel for 21 Nov., 1877. See also Girard, M. See same title in Section IIIf.

Conybeare, William Daniel, and Phillips, William. Outlines of the geology of England and Wales, with an introductory compendium of the general principles of that science and comparative views of the structure of foreign countries; illustrated by a coloured map and sections, etc. Part 1. [all publ.] 8°. London. 1822. pp. (8), 61, (1), 470, map, tables, pl. of sections.

Brief mention of the coleopterous remains in the calcareous slate of Stonesfield (oolite) will be found on pp. 207–209.

Fritsch, A. Palaeontologische untersuchungen der einzelnen schichten in der böhmischen kreideformation. (Archiv naturw. landesdurchf. Böhmen, 1, ii: 181–256, pl. 3.) 8°. Prag. 1869.

Refers on p. 187 to the discovery of an elytron of a beetle, and a tube of a phryganid larva in clay schists at Kounie.

—— Dva noví hrouci z českého útvarn křídového. (Vesmír, 18: 8, fig. 5.) 4°. [v. Praze.] 1888.

Describes and figures, much enlarged, two Coleoptera referred to the genera Lamiites and Velemnvakya. A brief list of the nine species recognized in the Bohemian chalk is added.

Geinitz, H. B. Charakteristik der schichten und petrefacten des sächsischen kreidegebirges. 4°. 1839–42. Dresden und Leipzig. pp. 4, 116, 26, pl. A, 24.

Under the head of Insecten, pp. 12–13, taf. 3–6, are described and figured borings of insects which the author, supported by Reichenbach and Germar, refers to Cerambyoidæ, and describes under the generic appellation Cerambycites. Dr. Geinitz now informs me that these belong to Gastrochæna amphisbæna Goldf., a burrowing mollusk.

George, H. See Brongniart, C. J. E.

Germar, E. F. See Geinitz, H. B.

G[irard], M. Les perforations des bois fossiles. (La nature, 6: 112, figs. 1–6.) 4°. Paris. 1878.

Popular account of Brongniart's two papers on the subject, with figures. See same title in Section III f.

Heer, O. Beschreibung der angeführten pflanzen und insekten. 4°. n. p., n. d. pp. 21, taf. (3). (Escher v. d. Linth, Geologische bemerkungen über das nördliche Voralberg und einige angrenzenden gegenden. pp. 115–135, taf. 6–8.) (Neue denkschr. allg. schweiz. gesellsch. gesammt. naturw., 13.) 4°. Zürich. 1853.

B. Insekten. pp. 18–21 (133–135), taf. 7. Describes two beetles from the trias of Vaduz.

—— Beiträge zur jura flora Ostsibiriens und des Amurlandes. t. p., pp. 122, pl. 31. (Mém. acad. imp. sc. St. Pétersb., (7), 22, xii.) 4°. St.-Pétersbourg. 1876.

Elaterites sibiricus is described on p. 41.

Forms vol. 4, no. B. of Heer's Flora fossilis arctica.

—— Ueber einige insektenreste aus der raetischen formation Schonens. (Förhandl. geol. fören. Stockh., 4, vii: 192–197, taf. 13.) 8°. Stockholm. 1878.

Description and figures of eight Coleoptera.

Heyden, C. von. Chrysobothris veterana und Blabera avita, zwei fossile insekten von Solenhofen. (Palaeontogr., 1: 99–101, pl. 12, fig. 4–5.) 4°. Cassel. 1817.

Simple description. See same title in Section V d.

Lartigue. Échantillons de bois fossiles provenant du gault de Lottinghen. (Ann. soc. ent. France, (5), 6, bull. ent., 107.) 8°. Paris. 1876.

Exhibition of fossil wood perforated by insects, afterward reported on by Brongniart, C. J. E. (q. v.)

Linné, C. von. Wästgöta resa på rikseus höglofliga ständers befallning förrättad år 1746. Med anmärkningar uti oeconomien, naturkunnogheten, antiquiteter, inwärnarnes seder och lefnads-sätt, med tilhöriga figurer. 16°. Stockholm. 1747. pp. (12), 284, (19), pl. 5.

Refers, on p. 24, to finding beetles in the limestone of Kinnekulle.

TRANSLATION: Reisen durch Westgothland, welche auf befehl der hochlöblichen reichsstände des königreichs Schweden im jahr 1746 angestellt worden. 8°. Halle. 1765. pp. (20), 318, pl. 7.

Not seen. The same (probably) is found on p. 30.

Mantell, G. A. Notes on the wealden strata of the Isle of Wight, with an account of the bones of iganodons and other reptiles discovered at Brook Point and Sandown Bay. (Quart. journ. geol. soc. Lond., 2, i : 91–96.) 8°. London. 1846.

In the closing paragraphs brief reference is made to elytra of "two or more species of Coleoptera" at Waterlogbory.

—— A brief notice of organic remains recently discovered in the wealden formation. (Quart. journ. geol. soc. Lond., 5: 37–43, pl. 3.) 8°. London. 1849.

A brief notice, p. 39, of elytra of Coleoptera from the "fresh-water strata above the oolite in Buckinghamshire;" two of these are figured in detail, but no suggestion is made of their affinities.

Moore, C. On the zones of the lower lias and the Avicula contorta zone. (Quart. journ. geol. soc. Lond., 17, i : 483–516, pl. 15–16.) 8°. London. 1861.

On p. 513 mentions the occurrence of Caralidæ at Vallis.

Phillips, W. See Conybeare, W. D., and Phillips, W.

Prévost, Constant. Observations sur les schistes calcaires oolitiques de Stonesfield en Angleterre, dans lesquels ont été trouvés plusieurs ossemens fossiles de mammifères. (Ann. sc. natur., 4 : 389–417, pl. 17–18.) 8°. Paris. 1824.

Refers merely, p. 393, and in explanation of plates, p. 417, to elytra of a Buprestis? figured pl. 18, fig. 26.

Reichenbach, Heinrich Gottlieb Ludwig. See Geinitz, H. B.

Roemer, F. Notiz über ein vorkommen von fossilen käfern (coleopteren) im rhät bei Hildesheim. (Zeitschr. deutsch. geol. gesellsch., 28: 350-353, figs.) 8°. Berlin. 1876.

Describes and figures three species of Coleoptera, of which one is not named.

Westwood, J. O. [Exhibition of a fossil beetle from Stonesfield.] (Trans. ent. soc. Lond., 4, journ. of proc., 40.) 8°. London. 1841.

Regards the elytra figured by Buckland (pl. 46°, figs. 4-9) as prionideous, not buprestideous.

Vg.—Mesozoic Diptera.

₊ See under Section I and Section IV, there being no separate titles under this head.

Vh.—Mesozoic Lepidoptera.

₊ See also under Section I, Section IV, and Section Ve.

Saporta, G. de. [Lettre à M. de Selys-Longchamps.] (Compt. rend. soc. ent. Belg., (2), xxiii: 8.) 8°. Bruxelles. 1876. (Ann. soc. ent. Belg., 19: 4.) 8°. Bruxelles. 1876.

Lepidoptera could not have appeared before phanerogamous flowers. See also Section IIId.

Vi.—Mesozoic Hymenoptera.

₊ See under Section I and Section IV, there being no separate titles under this head.

VI.—GENERAL FOR CENOZOIC TIME.

₊ See also under Section I.

Aldrovandus, U. Mvsaevm metallicvm in libros IIII distribvtvm. f°. Bononiae. 1648. t. p., ff. 2, pp. 979, (13).

Cap. 18, Do svccino, sev electro, pp. 403-418, refers briefly to insect inclusa on p. 406, under the side heading Quae animantes in svccino sint.

[Argenville, Ant. Jos. Desallier, d'.] L'histoire naturelle éclaircie dans une de ses parties principales, l'oryctologie, qui traite des terres, des pierres, des métaux, des minéraux et autres fossiles, ouvrage dans lequel on trouve une nouvelle méthode latine et françoise de les diviser, et une notice critique des principaux ouvrages qui ont paru sur ces matières. Enrichi de figures dessinées d'après nature. Par M. * * * des Sociétés royales des sciences de Londres et de Montpellier. 4°. Paris. 1755. pp. (8) 16,562, pl. 26.

Mentions some fossil insects on pp. 83,353; and on p. 360, pl. 21, refers to what is called a "chenille" and "deux papillons."

Assmann, A. Palaeontologie. Beiträge zur insekten-fauna der vorwelt.—Einleitung. I. Beitrag. Die fossilen insekten des tertiären (miocenen) thonlagers von Schossnitz bei Kanth in Schlesien. II. Beitrag. Fossile insekten aus der tertiären (oligocenen) braunkohle von Naumburg am Bober. Mit einer tafel abbildungen. 8°. Breslau. 1869. pp.

Assmann, A.—Continued.
1-62, taf. 1. (Zeitschr. entom. ver. schles. insektenk., (2) 1.),

The first paper contains an account of the geology and paleontology of Schossnitz with full descriptions of ten species of insects.

See same title in Section I and Section VII.

Aurifaber, Andr. Succini historia, oder Bericht woher der agt- oder hürnstein ursprünglich komme. 4°. Königsberg. 1551.

Not seen. Dr. Hagen informs me that it contains references to insects in amber, and is therefore one of the earliest works mentioning them. He thinks, indeed, that Münster's earlier mention of them may have been due to information received direct from Aurifaber; both were disciples of Luther. Later editions were published in 1557 and 1572 in 4°, and rendered into Latin verse by Scholzius, in 1503 and 1671 in 8°.

Aycke, Johann Christian. Fragmente zur naturgeschichte des bernsteins. 16°. Danzig. 1835. pp. 8, 107.

Contains a section: In bernstein eingeschlossene organische gegenstände, pp. 58-64, where the author states that insects occur most frequently in transparent amber and in such as shows a concentric structure; oftener also in amber brought up from the sea and only rarely in that dug from the earth. He also criticises various authors and especially Schweigger for confounding copal and other gums with amber.

Aymard, Auguste. La découverte d'un assez grand nombre d'insectes dans les

Aymard, A.—Continued.
marnes subordonnées à la formation gyp-
seuse près du Puy. (Bull. soc. géol. France,
6: 235–236. 8°. Paris. 1835.

A brief announcement with a consideration of
its geological import.

——— Rapport sur les collections de
M. Pichat-Dumazel. (Cong. scient. France,
sess, 22: 42.) 8°. Le Puy. 1854.

Not seen; quoted from Dustalet. Names, but
neither describes nor figures, seven insects from
Le Puy, of which three belong to Coleoptera,
three to Diptera and one to Neuroptera.

Ballenstedt [Johannes Georg Justus].
Entdeckung von insekten-nestern der urwelt im bernstein. (Archiv nenost. entd.
urw., 5: 28–40.) 8°. Quedlinburg und
Leipzig. 1823.

A very full abstract of Troost's paper, with
comments.

Bargagli, Piero. Di tre opuscoli sugli
insetti fossili e sulle formazioni inglesi e
straniori nelle quali sono stati scoperti
avanzo d'insetti, pubblicati da H. Goss
(Bull. soc. ent. ital., 12: 127–138.) 8°.
Firenze. 1880.

A very full abstract of the first of Goss's series
of three papers (q. v.).

Bassi, Carlo Agostino. Memorie intor-
no allo studio degl' insetti fossili in genere. (Atti reun. scienz. ital., 3: 400–401.)
4°. Firenze. 1841.

An account of three or four insects from Sini-
gaglia in the Milan museum, to only one of which
a name—Cleonolithus antiquus—is given. None
of them are properly described.

An abstract entitled "Ueber die wichtigkeit des
studiums der fossilen knochen [kerfen?] für die
geologie" will be found in Oken's Isis. 1843, pp.
418–419. 4°. Leipzig. 1843.

Beck [H.] Notes on the geology of Den-
mark. (Proc. geol. soc. Lond., 2: 217–
220.) 8°. London. 1836. (Lond. Edinb.
phil. mag., (3), 8: 553–556.) 8°. London.
1836.

Contains a paragraph relating to tertiary deposits of Jutland "older than the erratic blocks"
and containing "the elytra of beetles, the cases
of the larvae of Phrygauea, and a hymenopterous
insect which the author has called Cleptis steen-
strupii."

Bell, Alfred. Post-glacial insects (Entom., 21: 1–2). 8°. London. 1888.

Concludes with a list of Coleoptera 26 sp.,
Hemiptera 1, Diptera 1, and Neuroptera 2, from
peat in England.

Berendt, Georg Carl. Die insekten im
bernstein. Ein beitrag zur thiergo-

Berendt, G. C.—Continued.
schichte der vorwelt. 1er heft. 4°. Danzig. 1830. pp. (2), 39.

Only pp. 29–39 deal with the insects themselves,
and the remarks are of a very general nature, but
give the first published information concerning
amber insects based on considerable collections.
Hagen (Bibl. ent., 1: 42) records plates to a second
part.

——— Die im bernstein befindlichen
organischen reste der vorwelt, gesammelt
in verbindung mit mehreren bearbeitet
und herausgegeben von Dr. Georg Carl
Berendt. The covers entitled: Organische
reste im bernstein. 2 v. f°. Berlin.
1845–56.

Contains four parts; the first volume has the
plants by Goeppert and Berendt (with general
chapters on amber by Berendt), and the Crustacea, Myriapoda, Arachnida, and Aptera by Koch
and Berendt; the second volume, the Hemiptera
and Orthoptera by Germar and Berendt; and the
Neuroptera by Pictet and Hagen. (See these authors, the last named in Section VIIc.)

——— See also **Hope, F. W.**; and
Troost, G.

Beringer, J. D. A. Lithographiae
wirceburgensis, ducentis lapidum figuratorum, a potiori insectiformium, prodigi-
osis imaginibus exornatae specimen primum, quod in dissertatione inaugurali
physico-historica, cum connexis corolla-
riis medicis, authoritate et consensu in-
clytae facultatis medicae, in alma coo-
francica wirceburgensium universitate,
preside . . . D. Joanne Bartholomaeo
Adamo Beringer . . . exantlatis de more
rigidis examinibus pro suprema docto-
tus medici laurea, annexisque privilegiis
rité consequendis, publicae literarum
disquisitioni submittit Georgius Ludovicus Hueber, . . . In consueto auditorio
medico. Anno 1726. f°. Wirceburgi. pp.
(12), 96, tab. 21.

Nearly half of the plates contain grossly exag-
gerated, worthless figures of insects, the text for
which is crowded on p. 94 by mere descriptive titles to the plates. The specimens from which the
plates were drawn are said to have been fabrica-
tions. Weigel's sale catalogue, n. f. 33, for 1887, records a copy with 22 plates.

Bertkau, Ph. Ueber Planocephalus
aselloides Scudd., und Limnochares antiquus v. Heyl. (Sitzungsb. niederrh. gesellsch. natur- u. heilk., 1885, 298–300.)
8°. Bonn. 1885.

Regards these two insects as closely related and
as belonging to the Galgulidae. The article forms
pp. 4–5 of an untitled collection (6 pp.) of the author's papers from the Sitzungsberichte.

Bertrand, Élie. Dictionnaire universel des fossiles propres et des fossiles accidentels, contenant une description des terres, des sables, des sels, des soufres, des bitumens, des pierres simples & composées, communes & prétieuses, transparentes & opaques, amorphes & figurées, des minéraux, des métaux, des pétrifactions du règne animal & du règne végétal, &c., avec des recherches sur la formation de ces fossiles, sur leur origine, leur usage, &c. 2 v. 8°. La Haye. 1763.

Under the heading Entomolithes, vol. 1, pp. 201-202, is a very brief account of those then known, with bibliographical references.

Bleicher, Marie Gustave. Note sur la découverte d'un horizon fossilifère à poissons, insectes, plantes, dans le tongrien de la Haute-Alsace. (Bull. soc. géol. France, (3), 8: 222-229.) 8°. Paris. 1880.

Records, pp. 226-227, the occurrence at Rouffach of insects, referred to Cicadariae and Hymenoptera, and of an apterous articulate, probably an isopod crustacean, but perhaps a cockroach.

Bock, Friedrich Samuel. Versuch einer kurzen naturgeschichte des preussischen bernsteins und einer neuen wahrscheinlichen erklärung seines ursprunges. 16°, Königsberg. 1767. pp. 146.

Gives on pp. 138-146 a list of some insects and other animals found in amber.

—— Versuch einer wirthschaftlichen naturgeschichte von dem königreich Ost- und West-Preussen. 5 v. 16°. Dessau. 1782-'85.

Bd. 2 (1783) contains a short passage, pp. 106-107, on insects in amber.

Bolton, John. On a deposit with insects, leaves, etc., near Ulverston. (Quart. journ. geol. soc. Lond., 18, proc., 274-277, figs. 1-2, sections.) 8°. London. 1862.

Mentions only a few insects determined in a general way by Mr. Henry Tibbats Stainton.

Bonnac, de. Sur l'ambre jaune. (Hist. acad. roy. sc., 1705, 41-44.) 4°. Paris. 1730.

Concludes amber to be formed on the land from its inclosure of terrestrial animals "comme des mouches, des fourmis," etc.

Born, Ignace [Edl.] de. Catalogue méthodique et raisonné de la collection des fossiles de Mlle Éléonore de Raab. 2 v. 8°. Vienne. 1790. Tom. 1, pp. (42), 502;—tom. 2, pp. (8), 499, (66), tab 1.

Contains: Pétrifications d'insectes - entomolithes, vol. 1, pp. 464-466; mentions four insects.

Boué, A. Mémoire géologique sur le sol tertiaire et alluvial du pied septentrional des Alpes allemandes. [After part i entitled: Description du sol tertiaire, situé au pied des Alpes allemandes, et dans la Hongrie et la Transylvanie.] (Journ. de géol., 2: 333-385; 3: 1-35, 97-143, pl. 2.) 8°. Paris. 1830-'31.

Mentions, 3: 105, and figures, pl. 2, some insects of Radoboj.

[Braun, F.] Verzeichniss der in der kreis naturalien sammlung zu Bayreuth befindlichen petrefacten. 4°. Leipzig. 1840. pp. 8, 118, karte, tabelle, taf. 22.

Div. 6, Insecta, p. 71, mentions three insects from the braunkohl.

Brodie, P. B. Exploration of the leaf-beds in the lower Bagshot series of Hants and Dorset. (Geol. mag., 7: 141.) 8°. London. 1870.

Suggestions for further search.

—— The nature, origin, and geological history of amber, with an account of the fossils which it contains. 8°. [Warwick.] n. d. pp. 15. (Ann. rep. proc. Warw. nat. arch. field club, 1878?)

Notices of the insect inclosures are mostly confined to pp. 8-12, and are of a general nature; separate only seen.

—— On the discovery of a large and varied series of fossil insects and other associated fossils in the eocene (tertiary) strata of the Isle of Wight. 8°. Warwick. 1878. pp. 12. (Ann. rep. proc. Warw. nat. arch. field club, 1878.) 8°. Warwick. 1878.

A general popular account; only the separate paper seen.

Brongniart, Alexandre. Succin. (Dict. sc. nat., 51: 229-240.) 8°. Strasbourg et Paris. 1827.

Mentions in general terms (pp. 233-234) the insects most commonly found in amber.

Bruckmann, A. E. Flora oeningensis fossilis. Die oeninger steinbrüche, das sammeln in denselben und die bis jetzt dort gefundenen pflanzenreste. (Jahresh. ver. vaterl. naturk., 6: 215-238.) 8°. Stuttgart. 1850.

The introductory part, pp. 215-225, gives an account of the different beds of the Oeningen quarries and the animals as well as plants characteristic of each. Insects are mentioned on pp. 217 and 222. The appendix (ibid., 8: 252-254,—1852) refers only to plants.

Brydone, Patrick. A tour through Sicily and Malta, in a series of letters to

Brydone, P.—Continued.

William Beckford, Esq., of Somerly in Suffolk. A new edition. 2 vols. 8°. London. 1776. vol. 1, map, pp. 16, 373; —vol. 2, pp. 11, 355.

Vol. 1, pp. 282-284 contains a short account of the amber found at the mouth of the Giaretta containing "flies and other insects." The first edition was published in 1774; another in 1775; another with precisely the same pagination as in the one quoted, in 1790; another in 1806; a French translation in 1775 at Paris, and in 1819 at Frankfort; and a German translation in 1777 at Leipzig.

Burmeister, H. Kerfe der urwelt. (Burm., Handb. d. ent., 1 : 632-640.) 8°. Berlin. 1832.

A brief resumé of what was known at that time, with some original statements concerning the insects observed by him in amber.

TRANSLATION: Insects of a former world. (Burm.-Shuck., Man. ent., pp. 574-581.) 8°. London. 1836.

—— See also **Hünefeld, L.**

Buttner, David Sigismond. Rudera diluvii testes, i. e. zeichen und zeugen der sündfluth; in ansehung des itzigen zustandes unserer erd- und wasser-kugel; insonderheit der darinnen vielfältig auch zeither in querfurtischen revier unterschiedlich angetroffenen; ehemahls verschwemten thiere und gewächse; bey dem lichte natürlicher weissheit betrachtet; und nebst vielen abbildungen zum druck gegeben. 4°. Leipzig. 1710. front., t. p., pl., pp. (6), 314, (20), pl. 30.

Figures a few insects on plates 16 and 23, briefly referred to on page 226.

Capellini, Giovanni. Pesci ed insetti fossili nella formazione gessosa del Bolognese. (Gazz. dell' Emilia, no. 141.) 1869.

Not seen; quoted from Malfatti.

—— Il calcare di leitha, il sarmatiano e gli strati a congerie nei Monti di Livorno, di Castellina marittima, di Miemo, e di Monti Catini. Considerazioni geologiche e paleontologiche. (Atti accad. lincei, (3), mem. classe sc. fis., 2: 275-291.) 4°. Roma. 1878.

Gives on p. 285 a list of six insects found in the diatomaceous schists of Gabbro, two identical with Oeningen species.

—— Gli strati a congerie o la formazione gessosa-solfifera nella provincia di Pisa e nei dintorni di Livorno. (Atti

Capellini, G.—Continued.

accad. lincei, (3), mem. sc. fis., 5: 375-427, pl. 1-9.) 4°. Roma. 1880.

Refers in several places to the occurrence of insects, and especially of larvæ of Libellula in different strata and localities.

Chantre, Ernest. See **Lartet, L.**, et **Chantre, E.**

Charpentier, T. von. Beschreibung eines libellulinits aus Kroatien. (Neues jahrb. f. miner., 1841, 332-337, pl. 1.) 8°. Stuttgart. 1841.

Description, figure, and discussion of the affinities of a beautifully preserved dragon-fly's wing from Radoboj.

—— Ueber einige fossile insecten aus Radoboj in Croatien. (Nova acta acad. leop.-carol., 20, i: 399-410, tab. 21-23.) 4°. Vratislaviæ, etc. 1843.

Descriptions and (poor) illustrations of seven tertiary insects.

Cope, E. D. Report on the vertebrate paleontology of Colorado. (Ann. rep. U. S. geol. surv. terr., 1873, 427-533, pl. 1-8.) 8°. Washington. 1874.

On pp. 439-440 he refers to the horizon of the insect-bearing "Green-river shales," and records "insects and their larvæ"—some of the latter "nearly an inch long, and others minute and in prodigious numbers"—from Fontanelle Creek, and the "east side of Green River above the mouth of Labarge Creek."

Cope, E. D. A contribution to the knowledge of the ichthyological fauna of the Green River shales. (Bull. U. S. geol. surv. terr., 3: 807-819.) 8°. Washington. 1877.

He announces on p. 807 the discovery of fossil insects "near the mouth of Labarge Creek," Wyoming Territory, and at another locality "nearer the main line of the Wasatch Mountains."

Coquand, Henri. Sur la découverte faite dans les plâtrières d'Aix d'une grenouille fossile. (Bull. soc. géol. France, (2), 2: 383-386.) 8°. Paris. 1845.

Announces also the occurrence of insects in the same locality. And reviews what is known of them from the papers of Boisduval and Curtis.

A notice, not seen, will be found in Froriep's Neue notizen, 37: 33-36. 4°. Erfurt. 1846.

Cordus, Valerius. Sylva, qua rerum fossilium in Germania plurimorum metallorum, lapidum et stirpium variarum notitiam lapidum brevissime persequitur. f°. Tiguri. 1561.

Not seen; said to contain references to fossil insects from Oeningen.

Costa, Achille. See **Hope, F. W.**

Croizet (*l'abbé*). Indications des fossiles de diverses natures qu'on trouvé dans la montagne de Gergovie, avec les couches qui leur servent de gisement. (Bull. soc. géol. France, 7 : 104–106.) 8°. Paris. 1836.

Merely mentions the occurrence of insects at Morlague, p. 106. Noticed in Neues jahrb. mineral., 1836, 626–627. 8°. Stuttgart. 1826.

Curtis, John. Observations upon a collection of fossil insects discovered near Aix in Provence, in the summer of 1828, by R. J. Murchison and Charles Lyell, jr. (Murch.-Lyell, Tert. fresh water form. Aix, pp. 9–13, pl. 6.) 8°. Edinburgh. 1829. (Edinb. new. phil. journ., 7 : 293–297, pl. 6.) 8°. Edinburgh. 1829.

A list, with occasional brief descriptions, of forty-seven species; the first important notice of the Aix insects.

—— See also **Murchison, R. I.**; **Murchison, R. I., and Lyell, C.**

Dawson, George Mercer. Sketch of the geology of British Columbia. 8°. London. 1881. pp. 19. (Geol. mag., (2), 8 : 156–162, 214–227.) 8°. London. 1881.

Refers, p. 8 (162), to the general conclusions furnished by the tertiary insects of the interior plateau.

—— See also **Selwyn, A. R. C., and Dawson, G. M.**

Deichmüller, J. V. Fossile insecten aus den diatomeenschiefer von Kutschlin bei Bilin, Böhmen. 4°. Halle. 1881. pp. 39. pl. (Nova acta leop.-carol.-deutsch. akad. naturf., 42 : 293–331, tab. 21.) 4°. Halle. 1881.

Describes and figures thirteen species, all but one of them new, and mostly Coleoptera. They indicate a warmer climate at the time.

Desmarest, Anselme Gaetano. Insectes dans le succin.

Not seen. Cf. Marcel de Serres in Ann. sc. nat., 15 : 102. 8°. Paris. 1828. Also Guérin in Dict. class. hist. nat., 8 : 580. 8°. Paris. 1825.

Ehrenberg, Christian Gottfried. Eine sammlung bei Brandenburg aufgefundener bernsteinstücke. (Froriep. None notizen geb. nat.- heilk., 19 : 120.) 4°. Weimar. 1841.

Ehrenberg, C. G.—Continued.

Refers to insects of the genera Ceratopogon and Gryllus and to Phryganidae and nuts. Notice of same in Neues jahrb. mineral., 1842, 502. 8°. Stuttgart. 1843.

Erichson, W. F. See **Maravigna, C.**

Eser, Anton Friedrich. Das petrefactenlager bei Ober- und Unter-Kirchberg an der Iller im oberamt Laupheim. (Jahresh. ver. vaterl. naturk. Württ., 4 : 258–267.) 8. Stuttgart. 1849.

Records, pp. 264–265, the discovery of two insects " welche einige ähnlichkeit toit der wasserspinne (Argyroneta aquatica) und mit Cimex haben," in the miocene fish beds of Unterkirchberg.

Fischer (*of Mülhausen*). See **Foerster, B.**

Foerster, B. Mittheilungen über das oberelsässische tertiär. 8. Strassburg. 1885. pp. 9. (Tagebl. vers. deutsch. naturf., 58 : 386–394.) 4°. Strassburg. 1885.

Mentions half a dozen insects, mostly Coleoptera, found in the "marnes à cyrènes," nearly related to those of Oeningen.

—— Die gliederung des sundgauer tertiärs. (Mitth. comm. geol. Elsass-Lothr., 1 : 137–177, cuts, folding table.) 8°. Strassburg. 1888.

The insects on p. 165 were determined with the aid of Herr Fischer of Mülhausen.

—— Vorläufige mittheilung über die insekten des "plattigen steinmergels" von Brunstatt. (Mitth. comm. geol. Elsass-Lothr., 2 : 101–103.) 8°. Strassburg. 1889.

A bare list of genera recognized among the eighty-five species found; of these species, forty are Hemiptera, twenty-nine Coleoptera, and ten Diptera.

Fothergill, John. An extract of John Fothergill . . . his essay upon the origin of amber. (Phil. trans., 43 : 21–25.) 4°. London. 1746.

Mentions the occurrence in amber, of "ants, spiders, &c."

Fruglaye, de la. Extrait d'une lettre de M. de la Fruglaye à M. Gillet-Laumont, sur une forêt sous-marine qu'il a découverte près Morlaix (Finistère) en 1811. (Journ. des mines, 30 : 389–391.) 8°. Paris. 1811.

Notices a fossil chrysalis and a fly.

Gaudin, Charles Théophile. See **Heer, O.**

Geikie, James. Prehistoric Europe: a geological sketch. 8°. London and Philadelphia. 1881. pp. 18, 592, pl. 5.

Refers to notices of pleistocene insects on pp. 54, 256, 200, 440, 480, 494.

Geinitz, F. E. Siebenter beitrag zur geologie Mecklenburgs. 8°. Güstrow. 1885. 80 pp., 2 pl. (Arch. ver. freunde naturg. Meckl., 39: 41–120.) 8°. Güstrow, 1885.

Contains slight notices of peat insects.

Germar, E. F. Insecten in bernstein eingeschlossen, beschrieben aus dem academischen mineralien-cabinet zu Halle. (Germar, Mag. d. entom., 1: 11–18.) 8°. Halle, 1813.

Describes seven insects of various suborders. See also Schlechtendal, D. von.

—— Fauna insectorum Europæ. Fasciculus undevicesimus. Insectorum protogææ specimen, sistens insecta carbonum fossilium. Long, minute fol. Halæ. 1837. 2 t. p., 1 f., index, pl. 1–25.

Each plate contains one page of descriptive text, unpaged. The insects are all from the tertiaries.

—— Ueber einige insekten aus tertiärbildungen. (Zeitschr. deutsch. geol. gesellsch., 1: 52–66, taf. 2.) 8°. Berlin. 1849.

Describes and figures six beetles, a fly and a bee from the Rhine and Aix. Briefly noticed in Neues Jahrb. mineral., 1851, 759. 8°. Stuttgart. 1851.

Giard, A. See Saporta, G. de.

Giebel, C. G. Insectenreste aus den braunkohlenschichten bei Eisleben. (Zeitschr. gesammt. naturwissensch., 7: 384–386, taf. 5, figs. 1–4.) 8°. Berlin. 1856.

Describes and figures four insects,—two coleoptera, one cockroach, and one dragon-fly,—to only one of which, Buprestites minnæ, a name is given.

—— Wirbelthier und insektenreste im bernstein. (Zeitschr. gesammt. naturw., 20: 311–321.) 8°. Halle. 1862.

Describes sixteen insects of all orders.

—— Ueber eine kleine . . . suite von bernstein-insekten. (Zeitschr. gesammt. naturw., (2), 1: 87.) 8°. Berlin. 1870.

Mentions an ant and several genera of flies in a collection received from Dr. Schreiber.

Gilbert, Ludwig Wilhelm. See **Troost, G.**

Goebel, Severinus. De succino libr. II. quorum prior theologicus, posterior de succini origine agit. 8°. Francfurt. 1558.

Not seen; Dr. Hagen informs me that it contains references to insects in amber, among the earliest known.

Goeppert, H. R. On amber and on the organic remains found in it. (Quart. journ. geol. soc. Lond., 2, i: 102–103.) 8°. London. 1846.

A paragraph only on the insects, p. 102, specifying the orders and numbers of insects found.

—— Die tertiäre flora von Schossnitz in Schlesien. 4°. Görlitz. 1855. pp. 18, 52, pl. 26.

Six insects are figured on pl. 26, with a brief statement concerning them on p. vii.

—— Sull' ambra di Sicilia e sugli oggetti in essa rinchiusi. 4°. Roma. 1879. pp. 9, figs. (Mem. accad. linc., (3), mem. sc. fis., 3: 56–62.) 4°. Roma. 1879.

On p. 4 (58), he reviews the literature of the insects of Sicilian amber.

—— und **Berendt, George Carl.** Der bernstein und die in ihm befindlichen pflanzenreste der vorwelt. f°. Berlin. 1845. pp. (6), 4, 126, tab. 7. (Berendt, Org. reste bernst., bd. 1, abth. 1.)

Contains an important chapter, pp. 41–60, by Dr. Berendt, on Die organischen bernstein-einschlüsse im allgemeinen, treating of insects from p. 46 on, with many details and generalizations of interest, giving the first extended review of amber insects.

Goldfuss, Georg August. Beiträge zur kenntniss verschiedener reptilien der vorwelt. (Nova acta phys.-med. acad. leop.-car. nat. cur., 15, i (Verh. leop. car. akad. nat., 7, i): 61–128, pl. 7–13. 4°. Bonne. 1831.

On p. 118 will be found a list of genera of insecta found in the Rhenish braunkohl at Stösschen, Friesdorf, and Orsberg.

Goldsmith, E. On amber containing fossil insects. (Proc. acad. nat. sc. Philad., 1879, 207–208.) 8°. Philadelphia. 1879.

Mainly a description of its physical qualities, but mentions "ants, a fly, and probably small species of Coleoptera" in a fragment picked up on the shore of Nantucket, Mass.

Goss, H. Exhibition of a small collection of fossil insects obtained by Mr. J. S. Gardner from the Bournemouth leaf-beds (middle eocene). (Proc. entom. soc. Lond., 1878, 8.) 8°. London. 1878.

Merely mentions a few species by generic names.

Goss, H.—Continued.

—— Three papers on fossil insects, and the British and foreign formations in which insect remains have been detected. No. 1. The insect fauna of the recent and tertiary periods. 8°. [London, 1878.] pp. 65. (Proc. geol. assoc., 5, no. 6, pp. 282–343.) 8°. London, 1878.

ABSTRACT: The insect fauna of the tertiary period, and the British and foreign formations in which insect-remains have been detected. (Geol. mag. (n. s.), 4: 163–165.) 8°. London, 1877.

First read before the Brighton and Sussex natural history society; afterwards before the association. This abstract, and those of the succeeding papers of this series, were published previous to the full papers. See also Bargagli, P.

See same general title in Section II and Section IV, with note in former.

—— Introductory papers on fossil entomology. No. 9. Cainozoic time. On the insects of the eocene period, and the animals and plants with which they were correlated. (Entom. monthl. mag., 16: 124–128.) 8°. London, 1879.

The same. No. 10. Cainozoic time. On the insects of the miocene period and the animals and plants with which they were correlated. (Entom. monthl. mag., 16: 176–181.) 8°. London, 1880.

The same. No. 11. Cainozoic time. On the insects of the post tertiary or quaternary period and the animals and plants with which they were correlated. Entom. monthl. mag., 16: 198–201.) 8°. London, 1880.

See same title in Section I, Section II, and Section IV. See also Scudder, S. H., in Section I.

Gravenhorst, Johann Ludwig Christian. Bericht der entomologischen section. (Uebers. arb. veränd. schles. gesellsch. vaterl. cultur, 1834, 88–95.) 4°. Breslau. 1835.

On pp. 92–93 is given a list by genera of a collection of about 750 insects in amber exhibited before the entomological section of the society.

This communication has been frequently referred to Schilling, but apparently upon no proper grounds; his name does not appear.

—— See also Rathke, M. H.

Grew, Nehemiah. Musæum regalis societatis; or, a Catalogue and description of the natural and artificial rarities belonging to the Royal society and preserved at Gresham colledge; whereunto is subjoyned the comparative anatomy of

Grew, N.—Continued.
stomachs and guts. F°. London, 1681. pp. (12), 386, (2), (2), 43, portr., pl. 31.

On p. 341 (misprinted 334), he mentions amber containing cicadas, gnats, emmets, flies, and other insects. The addition of 1686 does not differ. Neither, apparently, do the editions of 1685 and 1694 which I have not seen.

Guérin-Méneville, F. E. See **Maravigna, C.**; and **Rondani, C.**

[**Hagen, H. A.**] Das bernsteinland. (Neue preuss. prov.-blätter, 10: 75–82, 120–125.) 16°. Königsberg. 1850.

A brief statement of the amber insects is given on pp. 124–125. The species are all extinct, the genera mostly still exist.

—— An entomological trip to Oxford. (Entom. weekly intell., 10: 165–168.) 8°. London, 1861.

Contains an account of the Sicilian amber in the Hope collection, with a notice of three species of white ants found therein.

—— Insekten im sizilianischen bernstein im oxforder museum. (Stett. entom. zeit., 23: 512–514.) 16°. Stettin. 1862.

More particularly concerned with a notice of three species of white ants, which is much the same as that given in the preceding.

—— On amber in North America. (Proc. Bost. soc. nat. hist., 16: 296–301.) 8°. Boston. 1874.

Recalls Troost's paper of 1821 on amber in Maryland and its reported inclusion of insect-nests; collects other references to American amber, and, pp. 300–301, refers to the resemblance of the fauna and flora of Prussian amber to that of the present time in North America, instancing the occurrence of Termopsis and Amphientomum among Neuroptera. See also Troost, G.

—— See also Hassencamp, E.; and **Krantz, A.**

Hagen, Karl Gottfried. Bemerkungen, die entstehung des bernsteins betreffend. (Beitr. kunde Preuss., 4: 207–227.) 16°. Königsberg. 1821.

Argues, p. 209, from the nature of the insects entombed in it, that amber is the gum of a tree.

Haidinger, Wilhelm. See **Heer, O.**

Hartmann, Philipp Jacob. Succini prussici physica et civilis historia cum demonstratione ex autopsia et intimiori rerum experientia deducta. 16°. Francofurti. 1677. front., pp. 291, pl. 3.

In book 1, chapt. 5, sect. 8, De inclusis, he mentions, p. 90, the occurrence in amber of "Aranearum non nunia species, insecta majores, minores; culices, crabrones, apes, tineas, blattas,

Hartmann, P. J.—Continued.

formicae, locustae;" and in book 2, chapt. 5, sect. 8, pp. 278-281, he endeavors to account for the occurrence of inclusions in general.

Another edition of the same date and place differs only in the plates, of which there are twenty according to Boehmer.

Noticed in Ephom. nat. cur., 3, 1 : 156.

———— Succincta succini prussici historia et demonstratio. 4°. Berolini. 1699. pp. (8), 48. (Phil. trans., 21 : 5-40.) 4°. London. 1699.

Sect. 3, c. iii : Animalculorum succino inclusorum accuratior demonstratio, pp. 19-31 (19-22 sep.), mentions in general terms the different sorts of insects known to be found in amber.

According to Boehmer (4 : 469) the separate edition was accompanied by six plates, but they do not exist in the two copies I have seen; the eight preliminary pages do not appear in the Philosophical transactions.

A very full abstract in English, under the title : An account of amber, appears in Phil. trans. abr., 2 : 473-493. 4°. London. 1749. The notice of the insects occurs on pp. 481-482. An abstract will also be found in Acta erud., 1700 : 332-336. 4°. Lipsiae. 1700.

Hassencamp, Ernst. Ueber fossile insekten der Rhön. (Würzb. naturwiss. zeitschr., 1 : 78-81.) 8°. Würzburg. 1860.

Contains MS. names of fossil insects by Heer, Hagen, and Heyden.

Hebenstreit, Johann Ernest. Mvsevm richterianvm continens fossilia animalia vegetabilia mar. illvstrata iconibvs & commentariis. Accedit de gemmis sculptis antiqvis liber singvlaris. 4°. Lipsiae. 1743. pp. 56, 384, (16), 34, portr., pl. 17.

A few fossil insects are specified on p. 256, and a "libella" figured pl. 13, fig. 2.

Heer, O. Ueber die fossile insektenfauna der tertiär-gebilde von Oeningen und Radoboj und die pflanzen aus gleicher formation an der hohen Rhone, aus einem briefe an Professor Bronn. (Neues jahrb. f. mineral., 1847, 161-167.) 8°. Stuttgart. 1847.

A catalogue of the beetles described in the first part of his classical work, with brief remarks on the general aspect of the fauna.

TRANSLATION: On the fossil insects of the tertiary formation of Oeningen and Radoboj. (Quart. journ. geol. soc. Lond., 3, ii : 60.) 8°. London. 1847.

Catalogue and concluding remarks omitted.

———— Physiognomie des fossilen Oeningen. 8°. [Winterthur. 1847.] pp. 22. (Vorhandl. schweiz. naturf. ge-

Heer, O.—Continued.

sellsch., 31 : 159-180.) 8°. Winterthur. 1847.

A general report on the insects will be found on pp. 167-174. Separate copy not seen.

———— Die insektenfauna der tertiärgebilde von Oeningen und von Radoboj in Croatien. Erster theil : Käfer. 4°. Leipzig. 1847. 1. p., pp. 2, 229, 1, pl. 8. (Neue denkschr. allg. schweiz. gesellsch. f. wissensch., 8.) 4°. Neuchâtel. 1847.

The same. Zweiter theil : Heuschrecken, fliegen, aderflügler, schmetterlinge und fliegen. Mit 17 lithographirten tafeln. 4°. Leipzig. 1849. pp. 264, 5, pl. 17. (Ibid., 11.) 4°. Zürich. 1850.

The same. Dritter theil : Rhynchoten. Mit 15 lithographirten tafeln. 4°. Leipzig. 1853. pp. 4, 138, pl. 15. (Ibid., 13.) 4°. Zürich. 1853.

This classical work is the most important ever published upon fossil insects, and may be called the first serious attempt at the classification of the tertiary species. Most of the material came from Oeningen and Radoboj, but it included all the author could examine from Aix and other localities. 462 species are described and figured, divided as follows: 119 Coleoptera, 39 Gymnognatha, 3 Neuroptera, 80 Hymenoptera, 9 Lepidoptera, 80 Diptera, and 133 Hemiptera. There are very few general observations, but attention should be called to an important excursus on the arrangement of the veins in the wings of insects and the elytra of Coleoptera, in the first part, pp. 76-95.

Reviewed by T. R. J[ones]. (Quart. journ. geol. soc. Lond., 9, ii, 35.) 8°. London. 1853.

Diagnoses of the species described in the third part appeared, without title, in the Bericht österr. litt. zool. bot. paleont., 1850-53, 193-203. 8°. Wien. 1855.

———— Fauna von Radoboj. Aus einem briefe [an Herrn bergrath Haidinger.] (Haidinger, Berichte, 5 : 86-87.) 8°. Wien. 1849.

Notice of the more remarkable insects belonging to the Vienna museum, and which are described in his great work, followed by brief remarks on the Radoboj insect fauna as a whole; and by comments of Haidinger.

———— Brief von O. Heer. (Haidinger, Berichte, 5 : 107.) 8°. Wien. 1849.

Notice of some peculiarities in the insect fauna of Radoboj in a letter to Unger.

———— Nachricht über die ersten ergebnisse einer durchsicht der reichen suite fossiler insecten, die von Herrn custos Freyer in Radoboj gesammelt . . .

Heer, O.—Continued.

worden waren. (Haidinger, Berichte, 6 :
5–7.) 8°. Wien. 1849.

Cursory report of his first examination of a
considerable collection of Radoboj insects, three-
fifths of which were found to be ants.

—— Die Morlot'sche sammlung von
fossilen insecten aus Radoboj. (Haidinger, Berichte, 6 : 132–134.) 8°. Wien.
1849.

A similar report to the last in the same volume;
the ants bear the same proportion as before and
make the tertiary European species equal in number to the living.

—— Flora tertiaria Helvetiæ. Die
tertiäre flora der Schweiz. 3 v. 4°. Winterthur. 1854–59. Bd. 1, 1854–55, t. p., pp.
6, 117, pl. 50 ; —bd. 2, 1856, pp. 4, 110, pl.
51–100 ; —bd. 3, 1859, pp. 6, 1–290, t. p.,
pp. 201–378, pl. 101–156, map.

Bd. 1 refers to Oeningen insects on pp. 10–11.
The latter half of bd. 3 was republished in 1860
under the title: Untersuchungen über das klima
u. s. w. (q. v.). See also the next entry, and Die
klimatischen verhältnisse, u. s. w., 1860.

—— Introduction à la flore tertiaire
de la Suisse, traduite par Charles Th.
Gaudin. (Arch. sc. phys. nat., 26 : 293–
314.) 8°. Genève. 1854.

A translation of the preliminary matter in the
first volume of the preceding: running references
to the insects of the period occur here and there,
especially on pp. 310, 311.

—— Ueber die fossilen insekten von
Aix in der Provence. (Vierteljahrsschr.
naturf. gesellsch. Zürich, 1 : 1–40, taf.
1–2.) 8°. Zürich. 1856.

The first important paper on Aix insects, cata-
loguing and describing sixty species of all orders,
preceded by remarks on the general characteristics of the fauna, which is considered to have
marked Mediterranean features.

—— Lettre à Sir Ch. Lyell [sur
l'étude de la flore tertiaire]. (Bull.
séances soc. vaud. sc. nat., 5 : 145–151, pl.)
8°. Lausanne. 1858.

American types among Oeningen insects, p. 148,
and relation of the Oeningen insects and plants,
p. 150.

—— Ueber die insectenfauna von
Radoboj. (Amtl. ber. vers. deutsch.
naturf., 32 : 116–126.) 4°. Wien. 1859.

A review of the subject based on the insects
described in his general work. The author finds
a commingling of European and Indian forms,
perfect dragon-flies but no larvæ, showing the
deposit to be marine; the occurrence of plants in
the same beds, with which the insects have special relations; a closer connection of Radoboj with
Aix than with Oeningen.

Heer, O.—Continued.

—— Die klimatischen verhältnisse
des tertiärlands aus O. Heer's tertiärflora
der Schweiz, bd. 3, s. 327–350, im aus-
zuge mitgetheilt. (Zeitschr. gesammt.
naturw., 15 : 1–12.) 8°. Berlin. 1860.

Insects are treated of on pp. 11, 12.

—— Untersuchungen über das klima
und die vegetationsverhältnisse des ter-
tiärlandes. Mit profilen und einem
kärtchen Europa's. Separatabdruck aus
dem dritten band der Tertiären flora der
Schweiz. 4°. Winterthur. 1860. t. p.,
pp. 170, pl. (156), karte.

Contains a couple of paragraphs, pp. 134–135
(334–335 of original) upon the tertiary insects and
the testimony they bear to the tropical and American nature of the time in which they lived. Another paragraph on pp. 60–61 (260–261) shows how
the condition of preservation of insects indicates
the season of their entombment.

TRANSLATION. Recherches sur le climat et la végétation du pays tertiaire.
Traduction de Charles Th. Gaudin. Avec
des profils et une carte de l'Europe. 4°.
Winterthur, Genève et Paris. 1861. pp.
220, 22, pl. 1, carte.

The paragraphs on pp. 134–135 of the original
are very much expanded on pp. 195–205 of this
translation, and include full tables of the families
of insects and their numerical representation in
the different European deposits of tertiary time.
Besides this, the Marquis Gaston de Saporta in
his included Examen des flores tertiaires de Provence, pp. 133–171, gives a paragraph, pp. 152–153,
concerning the insects of Aix. The remaining
paragraph referred to above appears unchanged
on p. 61.

—— On the fossil flora of Bovey
Tracey. (Phil. trans., 152 : 1039–1086, pl.
55–71.) 4°. London. 1862.

Insects from Bovey, p. 1082, pl. 58.

—— Flora fossilis arctica. Die fossile flora der polarländer enthaltend
die in Nordgrönland auf der Melville-
insel, im Banksland, am Mackenzie, in
Island und in Spitzbergen entdeckten
fossilen pflanzen. Mit einem anhang
über versteinerte hölzer der arctischen
zone von Dr. C. Cramer. 4°. Zürich.
1868. pp. 7, 192, map, pl. 50.

Contains: Fossile insecten von Nordgrönland,
pp. 129–130, pl. 19, 50; four species described.
Miocene flora von Island: Gliederthiere, pp. 154–
155, pl. 27; one beetle described.

Forms vol. 1 of Heer's Flora fossilis arctica.

—— Die miocene flora von Spitzbergen. Vorgetragen . . . bei der versammlung der schweizerischen natur-

Heer, O.—Continued.

forschenden gesellschaft, den 23 August,
1869, in Solothurn. 8°. Solothurn [n. d.].
pp. 15. (Verhandl. schweiz. naturf. gesellsch., 53: 156–168.) 8°. Solothurn.
1870. (Zeitschr. gesammt. naturw., (2),
1: 318–324.) 8°. Berlin. 1870.

Notices insects briefly at p. 12 (Verhandl. 166,
Zeitschr. 323).

TRANSLATION: La flore miocène du
Spitzberg. (Ann. sc. nat., (5), bot., 12:
302–311.) 8°. Paris. 1869.

Insects on pp. 303–309.

———— Preliminary report on the fossil plants collected by Mr. Whymper in
North Greenland in 1867. (Rep. Brit.
assoc. adv. sc., 39: 8–10.) 8°. London.
1870.

Two insects mentioned on p. 10, the same as in
the next.

———— Contributions to the fossil flora
of North Greenland, being a description
of the plants collected by Mr. Edward
Whymper during the summer of 1867.
(Phil. trans., 159: 445–488, pl. 39–56.) 4°.
London. 1870.

Contains description, pp. 484–485, and figures,
pl. 44, fig. 9, and pl. 56, fig. 14, of two insects. Cisto-
lites and Cercopidium, under the heading: Animals from Atanekerdluk. A. Insecta.

The paper forms vol. 2, no. i, of Heer's Flora
fossilis arctica.

———— Die miocene flora und fauna
Spitzbergens. Mit einem anhang über
die diluvialen ablagerungen Spitzbergens.
4°. Stockholm. 1870. pp. 98, taf. 16.
(Kongl. svenska vetensk.-akad. handl.,
8, vii.) 4°. Stockholm. 1870.

Zweiter abschnitt: Beschreibung der miocenen
thiere Spitzbergens. I. Insekten, pp. 73–78, pl.
16; contains descriptions of twenty-three insects,
of which twenty are Coleoptera.

Forms vol. 2, no. iii, of Heer's Flora fossilis
arctica.

———— Ueber die braunkohlen-flora
des Zsilythales in Siebenbürgen. 8°.
Pest. 1872. pp. 24, pl. 6. (Mitth. jahrb.
ung. geol. anst., 2, i.) 8°. Pest. 1872.

Mentions the discovery of fossil insects in the
tertiary beds of Tállya.

———— See also **Hassencamp, E.**

Helwing, Georg Andreas. Lithographia angerburgica, sive lapidum et
fossilium, in districtu angerburgensi &
ejus vicinia, ad trium vel quatuor millia-
rium spatium, in montibus, agris, arenofo-
dinis & in primis circa lacuum littora &
fluviorum ripas, collectorum brevis &

Helwing, G. A.—Continued.

succincta consideratio additis rariorum
aliquot figurit aeri incisis, cum praefa-
tione autoris & indicibus necessariis. 4°.
Regimonti. 1717. pp. (14), 96, (13), front.,
pl. 11.

The same. Pars II. In qva de lapidibvs
figvratis ad triplex regnvm minorale,
vegetabile et animalo redactis aliisqvo
fossilibvs in districtv angerbvrgensi
ejvsqve vicinia noviter detectis, et in
specio de origine lapidvm literas experi-
mentivm, occasione lapidis cvjvsdam ro-
saviensis, literas latinas L. V. R. repraе-
sentantis, svccincte disseritvr; additis
iconibvs rariorum. 4°. Lipsiae. 1720.
pp. 132, pl. 6.

On p. 78 is given a short notice of insects (formicae, blattae, tipulae, millipedes aliaqvo insecta) in
amber.

Henkel, Johannes Fridericus. De succino fossili in Saxonia electorali. (Acta
phys. med. acad. leop-carol., 4: 313–316.)
4°. Norimbergae. 1737.

Contains reference to insect inclusions on p.
316. Also said to be given in his Kleine miner.
chym. stud., p. 532. 8°. Dresden und Leipzig.
1741; the latter not seen.

Hensche, August. Ueber den bestand
und die neueren erwerbungen der born-
steinsammlung. (Schrift. phys.-ökon.
gesellsch. Königsb., 5, sitzungsb., 14–15.)
4°. Königsberg. 1864.

History of the growth and present extent of the
collection, rich in insect inclosures.

———— Bericht über die bernstein-
sammlung der königl. physikalisch-
ökonomischen gesellschaft. (Schrift.
phys.-ökon. gesellsch. Königsb., 6: 210–
215.) 4°. Königsberg. 1865.

Contains 853 specimens with insect inclosures,
of which over 6000 are Diptera; tables of the different groups are given on pp. 211–213.

Henslow, John Stevens. Supplementary observations to Dr. Berger's account
of the Isle of Man. (Trans. geol. soc.
Lond., 5: 482–505.) 4°. London. 1821.

Under the head of diluvial deposits, he refers,
p. 501, to a bed of peat in the parish of Kirk
Balaß, "containing a vast number of the exuviæ
of beetles, bees and their nests, crushed together
with sand vessels, rotten, but having their exter-
nal coating well preserved. . . . In general
the hard wings are the only parts of the beetles
which are preserved, and these are in appearance
as fresh as on a living insect. Dr. Leach was
enabled to identify a few with species at present
existing in England."

Heyden, C. von. Reste von insekten aus der braunkohle von Salzhausen und Westerburg. (Palæontogr., 4 : 198–201, pl. 37–38.) 4°. Cassel. 1856.

Divided into : Dicerca t-schef Heyden aus der braunkohle von Salzhausen, pp. 198-199, pl. 37, figs. 1-4.—Gänge von insekten-larven in hölzern der braunkohle von Salzhausen, pp. 199-200, pl. 38; borings of an Anobium, a Priouus, and a buprestid.—Fliegen aus der braunkohle der grube Wilhelmsfand bei Westerburg in herzogthum Nassau, pp. 200-201, pl. 37, figs. 6-8; three species described.

―――― Fossile insekten aus der braunkohle von Sioblos. (Palæontogr., 5 : 115–120, pl. 23.) 4°. Cassel. 1858.

Description of ten species, mostly beetles.

―――― Fossile insekten aus der rheinischen braunkohle. (Palæontogr., 8 : 1–15, pl. 1, 2, figs. 1–13.) 4°. Cassel. 1859.

Description of twenty-five insects of various orders.

―――― Gliederthiere aus der braunkohle des Niederrhein's, der Wetterau und der Rhön. (Palæontogr., 10 : 62–82, pl. 10.) 4°. Cassel. 1862.

Description of a crustacean, two arachnids, and thirty-two hexapods of various orders.

―――― See also Hassencamp, E. ; Meyer, C. H. von ; and Rathke, M. H.

Heyden, Carl von und Lucas von. Fossile insekten aus der braunkohle von Salzhausen. (Palæontogr., 14 : 31–35, pl. 9, figs. 13–22.) 4°. [Cassel.] 1865.

Description of twelve insects, mostly Coleoptera, and remarks on three others.

―――― See also Krantz, A.

Hope, Frederic William. Observations on succinic insects. (Trans. ent. soc. Lond., 1 : 133–147; 2 : 46–57, pl. 7.) 8°. London. 1836–37.

General remarks on the insects found in amber and gum anime, followed by a list of insects hitherto noticed by the author or known to Berendt. The species are all claimed as distinct from the recent, and to be tropical in their affinities.

―――― Observations on the fossil insects of Aix in Provence, with descriptions and figures of three species. (Trans. ent. soc. Lond., 4 : 250–255, pl. 19, figs. 1–3.) 8°. London. 1847.

Contains a list of genera occurring at Aix and "descriptions of three fossil species of insects" (Haluulaus, Rhynchaenus, Corizus) from the same locality.

Hope, F. W. Descrizione di alcune specie d' insetti fossili pel Rev. G. F. Hope; memoria presentata all' Accademia degli aspiranti naturalisti, in Dicembre 1847, ed inserita negli annali della stessa [with notes by A. Costa]. 8°. (Napoli.) n. d. pp. 7, pl. (Ann. acc. aspir. nat. Napoli, 1847. pp. —, tav. 10.) 8°. Napoli.

Five species described and figured.

Hueber, Georgius Ludovicus. See Beringer, J. D. A.

Hünefeld, L. Ueber bernstein-insecten. (Oken's Isis, 1831, 2000 [1100].) 4°. Leipzig. 1831.

A list of insects is given by Burmeister.

Ittiologia veronese del museo Bozziano ora annessa a quello del conte Giovambattista Gazola e di altri gabinetti di fossili veronesi con la versione latina. f°. Verona. 1796. pp. 52, 323, pl. 76.

Part 1, § 27, p. 31, records in the Boza museum, "duo Asili, Cimex pnus Americanus, omnes inde effossi."

John, Johann Friedrich. Naturgeschichte des suuccins, oder des sogenannten bernsteins; nebst theorie der bildung aller fossilen, bituminösen, inflammabilien des organischen reichs und den analysen derselben. 2 th. 16°. Köln. 1816. 1er th.. pp. 18, 438;—2er th., pp. 6, 125, (21).

A list of insects found in amber, arranged by genera, will be found in 1, pp. 221-223; and in 1, pp. 169-170, a bibliography of amber literature.

Jokély, Johann. Die tertiären süsswassergebilde des Egerlandes und der Falkenauer gegend in Böhmen (Jahrb. k. k. geol. reichsanst., 8 : 466–515.) 8°. Wien. 1857.

Remains of insects are recorded from Eger (p. 477), Krottensee (p. 482), Grasseth (p. 502).

Jones, Thomas Rupert. Fossil insects. (Geol. mag., 7 : 348.) 8°. London. 1870.

Correction of geological horizon of certain insects described by Westwood.

―――― See also Heer, O.

Karg, Joseph Maximilian. Ueber den steinbruch zu Ooningen bei Stein am Rheine und dessen petrefacte. (Denkschr. vaterl. gesellsch. aertze u. naturf. Schwabens, 1 : 1–73.) 8°. Tübingen. 1805.

Not seen. It contains references to the insects, and is mentioned by Heer.

Kawall, H. Der bernsteinsee in Kurland. (Correspondenzbl. naturf. ver. Riga, 6: 69–71. 8°. Riga. 1853.

Not seen; said to contain something on amber insects.

Knorr, Georg Wolfgang. Lapides, ex celeberrimorum virorum sententia diluvii universalis testes, quos in ordines ac species distribuit, suis coloribus exprimit, aeris incisos in lucem mittit et alia naturae miranda addit. *Also entitled:* Sammlung von merkwürdigkeiten der natur und den alterthümern des erdbodens, zum beweis einer allgemeinen sündfluth nach der meynung der berühmtesten maenner aus dem reiche der steine gewiesen und nach ihren wesentlichen arthen, eigenschaften, und ansehen, mit farben ausgedruckt, und in kupffer herausgegeben, in Nürnberg 1750. *With second title:* Sammlung von merckwürdigkeiten der natur und alterthümern des erdbodens welche petrificirte körper enthält ausgewiesen und beschrieben (erster theil). f°. Nürnberg. 1755. 2 t. p., pp. (2), 32, t. p. to atlas, tab. 1–38 (= 57 pl.).

19. 33 contains six figures, five of insects from Oeningen, the only distinguishable ones being three of odonate larvæ, explained on p. 27. To this work is appended, pp. 29–32, a letter from Myllus to von Haller, entitled Beschreibung einer neuen grönländischen thierpflanze. Bound up with the same is the following:

——— Die naturgeschichte der versteinerungen zur erläuterung der knorrischen sammlung von merkwürdigkeiten der natur, herausgegeben von Johann Ernst Immanuel Walch. Erster theil. f°. Nürnberg. 1773. pp. (6), 187.

This contains a further explanation of the plate on p. 181, in which the insects are called libellen, and which is preceded by an account (pp. 171–180) of what was then known of fossil insects, entitled Die entomolithen und helmintholithen.

Koch, (Friedrich) Carl Ludwig, und **Berendt** (Georg Carl). Die im bernstein befindlichen crustaceen, myriapoden, arachniden und apteren der vorwelt. f°. Berlin. 1854. t. p., pp. 4, 124, pl. 17. (Berendt, Bernst. befindl. org. reste vorw., 1, i.)

Edited with additions of importance by Menge. 10 Myriapoda, 123 Arachnida, and 21 Thysanura are described and figured, besides numerous others briefly described in the notes which Menge adds to nearly every species, nearly or

Koch und **Berendt,** etc.—Continued. quite doubling the extent of the text. Menge adds on pp. 7–8 a list of the species in his collection. Plates 1, 2, and 15 were different in earlier impressions. Plates 16 and 17 are supplementary.

Kollar, Vincent. See **Reuss,** A. E.

[**Krantz,** August.] Verzeichniss der von Dr. Krantz gesammelten, von Herrn Senator v. Heyden und Herrn Hauptmann v. Heyden in Frankfurt a. M. und von Herrn Dr. Hagen in Königsberg in der Palæontographica bis jetzt beschriebenen und abgebildeten insecten, etc., aus dem braunkohlengebirge von Rott im Siebengebirge. (Verhandl. naturh. ver. preuss. Rheinl. u. Westph., 24: 313–316.) 8°. Bonn. 1867.

Enumerates 73 Coleopt., 26 Dipt., 11 Neuropt., 4 Hymenopt., 3 Arachn., 2 Hemipt., 1 Lepidopt., 1 Orthopt. 120 species.

Lartet, Louis, et **Chantre,** Ernest. Études paléontologiques dans le bassin du Rhône; période quaternaire. (Arch. mus. hist. nat. Lyon, 1: 59–130.) 4°. Lyon. 1876.

Mentions the occurrence of insects at La Boisse, p. 104, and Sonnaz, p. 105.

Leach, William Elford. See **Henslow,** J. S.

Lindner. See **Schöberlin,** E.

Lyell, C. See **Murchison,** R. I., and **Lyell,** C.

MacCulloch, John. On animals preserved in amber, with remarks on the nature and origin of that substance. (Quart. journ. sc. lit. arts, 16: 41–48.) 8°. London. 1823. (Froriep, Notizen, 6, no. 114, pp. 49–51.) 4°. Erfurt. 1823.

Mainly devoted to describing the methods of distinguishing amber from other gums; insects and other animals are only mentioned in a general way.

Malfatti, Giovanni. Osservazioni sopra alcuni insetti fossili dell' ambra e del copale. 8°. Milano. 1878. pp. 15. (Atti soc. ital. sc. nat., 21: 181–195.) 8°. Milano. 1878.

Of a general nature, but contains at the close a list of additions to the Museo civico with remarks; and three pages of bibliography are appended.

——— Bibliographia degli insetti fossili italiani finora conosciuti. 8°. Milano. 1881. t. p., pp. 12. (Atti soc. ital. sc. nat., 24: 89–100.) 8°. Milano. 1881.

Malfatti. G.— Continued.

A valuable résumé of what has been published concerning the fossil insects of Italy, arranged by deposits. None are older than the tertiaries. Mention is made of three or four specimens in Italian museums, not before published.

Maravigna, Carmelo. Insectes dans l'ambre (Rev. zool., 1 : 162–169). [followed by remarks of Guérin Méneville] (pp. 169–170, pl. 1). 8°. Paris. 1838.

Maravigna's note is upon the conditions of occurrence of Sicilian amber. Guérin figures, enumerates and occasionally names about fifteen species, mostly Coleoptera, Hymenoptera, and Diptera.

Reviewed by Erichson with original notes in Ber. wiss. leist. entom., 1838, 29. 8°. Berlin. 1840. (Arch. f. naturg., 5, ii : 300.) 8°. Berlin. 1839. [1840.]

——— See also **Rondani, C.**

Massalongo, Abramo Bartolommeo Pr. Sopra un nuovo genere di pandance fossili della provincia veronese. 8°. Verona. 1853. pp. 16, (7, [201–207]), tav. 4. (Mem. accad. agric. Ver., 29 : 187–207, pl. 1–4.) 8°. Verona. 1854.

Refers an p. 12 (196) to the occurrence of Neuroptera, similar in form and size to the living Libellula, with "alcune piccole api" at Monte Bolca. The only copy I have seen is that of the separate paper, in which pp. 17–end are replaced by those of the academy's memoir.

——— Monografia delle nereide fossili. 8°. Verona. 1855. pp. 35, pl. 6.

In an appendix, pp. 31–32, he enumerates five species of insects from Monte Bolca.

——— Prodromo di un' entomologia fossile del M. Bolca. (Studii paleont., pp. 11–21, tab. 1 (pars.), 2.) 8°. Verona. 1856.

Describes seven insects of different orders.

Omboni says of this paper "che fa parte del suol studj paleontologici, pubblicati nel 1850 a Verona, nel Programma dell' I. R. Ginnasio liceale di quella città (tipografia Antonelli)."

——— Compendium faunae et florae bolcensis.

Not seen; nor have I been able to find a single reference to it in bibliographies, and presume it is still unpublished. It is referred to as above in several places by the author in other publications; see Studii paleont., p. 14, etc. It is not mentioned in Sordello's Bibl. paleont. vegt. ital. (1881.) Malfatti says of it (Bibl., 12) "rimasto inedito."

——— e **Scarabelli**, Giuseppe. Studii sulla flora fossile e geologia stratigrafica del Senigalliese. f°. Imola. 1859. pp. 8, 506, map, pl. 45.

Parte 1ª, Geologia stratigrafica is by Scarabelli; parte 2ª, Flora fossile by Massalongo. Insetti oo p. 25, contains a nominal list of species.

Massmann. See **Schöberlin, E.**

Matheron, Philippe. Compte rendu de la visite du terrain à gypse à Aix et du volcan de Beaulieu. (Bull. soc. géol. France, 13 : 451–465.) 8°. Paris. 1842.

The insects of the beds at Aix are referred to in general terms on p. 454, and their relative position pointed out.

——— Recherches comparatives sur les dépôts fluvio-lacustres tertiaires des environs de Montpellier, de l'Aude et de la Provence. 8°. Marseille. 1862. pp. 108 (?). (Mém. soc. émul. Marseille, 1 : 173–280.) 8°. Marseille. 1861.

Not seen; gives, according to Oustalet, some notice of Aix insects.

Meigen, Johann Wilhelm. See **Rathke, M. H.**

Meinecke, Johann Christoph. Vermischte anmerkungen über verschiedene gegenstände, sonderlich des steinreichs. (Naturforscher, st. 20 : 185–210.) 16°. Halle. 1784.

The first part, pp. 186–189, is devoted to amber as occurring in which he mentions various insects and Gryllus domesticus in particular.

Menge, A. Lebenszeichen vorweltlicher, im bernstein eingeschlossener thiere. (Progr. petrischule Danzig, 1856, pp. 1–32.) 4°. Danzig. [1856.]

Contains a valuable systematic review of the species in the author's collection, with occasional brief descriptions. The collection is one of the largest ever made, containing 67 Myriapoda of 31 species, 674 Arachnida of more than 150 species, and 3,102 Insecta, of which even the genera are rarely enumerated, but only separated by families.

——— Ueber ein rhipidopteron und einige andere im bernstein eingeschlossene thiere. Also entitled: Ueber ein rhipidopteron und einige helminthen im bernstein. 8°. Danzig. 1866. t. p., pp. 8, figs. (Schriften naturf. gesellsch. Danzig, (2), 1, iii–iv.) 8°. Danzig. 1866.

The strepsipteron is described and figured under the name of Triena tertiaria, and figures are given of a Chironomus to which a Mermis is found attached.

——— See also **Koch, F. C. L.,** und **Berendt, G. C.**

Mercati, Michael. Michaelis Mercati samminiatensis Metallotheca opus posthumum, auctoritate & munificentia Clementis XI. pontificis maximi e tenebris in lucem eductum; opera autem, & studio Joannis Mariae Lancisii archiatri

Mercati, M.—Continued.
pontificii illustratum. f°. Romae. 1717.
pp. 64, 378, (18), pl. 6, figs.

Atm. 5, cap. 2: De soccino, pp. 87-90, contains on p. 89 figures of some half dozen insects in amber which are enumerated in a single line on p. 88.

Meyer, C. E. H. von. Mittheilung an professor Bronn. (Neues jahrb. mineral., 20: 465-468.) 8°. Stuttgart. 1852.

Notices, p. 467, Heyden's Diecrea taschel: his discovery of insect borings in wood from the brown coal; and his statement that Xylophagus antiquus is a Bibio.

Millar, George Henry, *editor.* A new, complete, and universal body or system of natural history; being a grand, accurate, and extensive display of animated nature . . . written by a society of gentlemen. f°. London. n. d.

Not seen; according to Dr. Hagen the work mentions, p. 421, the presence of insects in amber.

Miller, S. A. The cenozoic or tertiary period. (Journ. Cinc. soc. nat. hist., 4: 93-144.) 8°. Cincinnati. 1881.

Gives lists of the tertiary insects described from North America.

Moore, C. Notes on a plant and insect bed on the Rocky River, New South Wales. (Quart. journ. geol. soc. Lond., 26: 261-263, pl. 18, figs. 2-11.) 8°. London. 1870.

Brief notice of the insects, about ten species of beetles, which are figured. An abstract with the same title (excepting the omission of the word Notes) on p. 2 of same, with discussion, which does not touch the insects. An abstract will also be found in Phil. mag., 39: 463. 8°. London. 1870.

Motschoulsky, Victor. Lettre à M. Ménétriés. (Études entom., 5: 3-38, pl.) 16°. Helsingfors. 1856.

Contains, pp. 25-30, a notice of Menge's collection of amber insects with descriptions, pp. 25-27, and figures (in the single plate) of thirteen Coleoptera. Also, p. 34, a brief notice of Heer's collection of Oeningen insects.

Münster, Sebastian. Cosmographiae universalis Lib. VI. in quibus, iuxta certioris fidei scriptorum traditionem describuntur, Omniñ habitabilis orbis partiñ situs, proprieaq dotes. Regionum Topographicae effigies. Terrae ingenia, quibus sit ut tam differentes & uarias species, & animatas & inanimatas, ferat. Animalium peregrinorum naturae & picturae. Nobiliorum ciuitatum icones & descriptiones. Regnorum initia, incrementa & translationes. Omnium gentium

Münster, S.—Continued.
mores, leges, religio, res gestae, mutationes: Item regum & principum genealogiae. Autore Sebast. Munstero. f°. Basileae. 1554. ff. (12), pl. (14), pp. 1163, + 1 folding plate not paged.

Liber III contains, pp. 783-784, a section De succino quod in Prussia legitur, in which, p. 784, amber is said to contain "hesticinæ, ut muscae, culices, apes, formicae, locustae, etc."

The German edition of 1598 contains this reference on pp. 1145-1146, and that of 1628 (?) on p. 1267.

Dr. Hagen has called my attention to this reference as perhaps the earliest mention of insects in amber. The same reference is doubtless contained in the latin edition of 1550, perhaps in the German edition of 1544. See Harv. univ. bull., 2: 285.

Murchison, R. I. On a fossil fox found at Oeningen near Constance; with an account of the deposit in which it was imbedded. (Trans. geol. soc. Lond., (2), 3: 277-290, pl. 33-34.) 4°. London. 1832.

The insects, pp. 286-287, pl. 34, are described by Curtis, with a mention of others determined by Samouelle; only a dozen species in all are mentioned; those figured are Odonata.

—— **and Lyell, Sir Charles.** On the tertiary fresh-water formations of Aix in Provence, including the coal field of Fuveau . . . with a description of fossil insects contained therein by John Curtis. (Proc. geol. soc. London, 1: 150-151.) 8°. London. 1829.

Accompanied, p. 151, by "Observations on the fossil insects" mentioned, by John Curtis, afterwards described and figured by the same. See Curtis, J.

Novák, O. Fauna der cyprisschiefer des egerer tertiärbeckens. Wien. 8°. 1877. pp. 26, pl. 3. (Sitzungsb. akad. wiss. Wien, 76, abth. 1, math. nat. classe, 71-96, taf. 1-3.) 8°. Wien. 1878.

Descriptions and illustrations of nineteen insects of various groups, but mostly Diptera.

Oken, Lorenz. Einige wörter über den Oeningensteinbruch. (Oken's Isis, 1840, 282-283.) 4°. Leipzig. 1840.

Refers to its fossil insects and particularly the larvae of Odonata.

Omboni, Giovanni. Di alcuni insetti fossili del Veneto. 8°. Venezia, 1886. pp. 16, pl. 2-3. (Atti r. ist. Ven. sc. lett. arti, (6), 4: 1421-1436, pl. 2-3.) 8°. Venezia. 1886.

Figures and describes a few species from Monte Bolca described previously by Heer and Massalongo and adds descriptions and figures of two Diptera and Coleoptera from Chiavon, Bolca, and Novale.

Oustalet, É. Recherches sur les insectes fossiles des terrains tertiaires de la France; première partie. Insectes fossiles de l'Auvergne. (Ann. sc. geol., 2, art. 3, pp. 1-178, pl. 1-6.) 8°. Paris. 1870.

—— The same, entitled: Mémoire sur les insectes fossiles des terrains tertiaires de la France. (Bibl. école hautes études, sect. sc. nat., art. 7, pp. 1-178, pl. 1-6.) 8°. Paris. 1871.

Studied descriptions of forty-five species, most of them new; they are mostly Diptera, especially Protomyine and Bibiones, and Coleoptera, especially Rhynchophora. The characteristics of the groups to which the insects belong are given in detail, and references made to other fossil insects of the same groups. The whole is preceded by a chapter of 43 pages, containing a good history of our knowledge of fossil insects, and is followed by one of 17 pages of general results reached by a study of the Auvergne fossil insects, which are found to show a mingling of indigenous and of exotic forms, a Mediterranean and American aspect, and a warmer climate than now.

An extended notice will be found in the Revue scient. France, (2), 4: 136-137. 4°. Paris. 1874. See also Giard., A., in Section VII.

—— Insectes de l'ambre. (Bull. soc. philom. Paris, (6,) 10: 98-99.) 8°. Paris. 1873.

A brief notice of various insects inclosed in a single block of amber in which Vaillant had detected a reptile. The insects are mentioned still more briefly in Vaillant's paper.

—— Les insectes fossiles de la France. (La nature, 3: 33-36, figs.) 4°. Paris. 1874.

A popular account of the tertiary insects of France, with figures of Biblo edwardsi, Colosoma agassizi and Cyllo sepulta.

—— Observation sur la communication de M. Filhol [sur les vertébrés fossiles des dépôts de phosphate de chaux du Quercy]. (Bull. soc. philom. Paris, (6,) 11: 21.) 8°. Paris. 1874 [1877].

Analogies between the insect fauna of the oligocene of southern France on the one hand, and that of southern N. America at the present time, or of the oligocene of the Rocky Mountains on the other.

Peale, Albert Charles. Report on the geology of the Green River district. (Ann. rep. U. S. geol. surv. terr., 1877, 511-646, pl. 47-76.) 8°. Washington. 1879.

Contains, p. 535, notice of the discovery of insects on Twin Creek, Wyoming Terr.; and pp. 633-839 a reprint of Scudder's description of Iulasia calculosa from Horse Creek Valley, Wyoming; see also p. 528.

Pictet de la Rive, F. J. Considérations générales sur les débris organiques qui ont été trouvés dans l'ambre et en particulier sur les insectes. (Arch. sc. phys. nat., 2: 5-16.) 8°. Genève et Paris. 1846.

Amber insects altogether differ specifically from living forms; a considerable number of genera are also distinct and there are two peculiar families: Archacides in the Arachnida and Pseudoperlides in the Neuroptera; a warmer climate than the present is indicated.

TRANSLATION: General considerations on the organic remains, and in particular on the insects, which have been found in amber. (Edinb. new phil. journ., 41: 391-401.) 8°. Edinburgh. 1846.

Presl, Joannes Swatopluceus. Additamenta ad Faunam protogaeam, sistens descriptiones aliquot animalium in succino inclusorum. (Deliciae pragenses, 1: 191-210.) 16°. Pragae. 1822.

Describes 1 species of Cynips, 6 Formica, 1 Tinea, 3 Tipula, 4 Musca, 2 Aranea, 1 Phalangium, and 1 Acarus. Noticed in Oken's Isis, 1823, 374-375. 4°. Jena. 1823.

Procaccini-Ricci, Vito. Lettera prima . . . sull'anatomia delle filliti sinigalliesi. (Nuov. ann. sc. nat., 1: 190-213, pl. 4-5.) 8°. Bologna. 1838.

Refers, p. 210, to a fossil insect which is figured with a leaf, pl. 5, fig. 1; no details are given and the figure is unrecognizable.

—— Lettera . . ., sugli entomoliti delle gessaje sinigagliesi. (Nuov. ann. sc. natur., ann. 4, tom. 7, pp. 448-456.) 8°. Bologna. 1842.

A general paper, in which on p. 449 it is stated that Coleoptera, Hemiptera, Lepidoptera, Neuroptera, Hymenoptera, Diptera, and Aptera have been found at Senigaglia.

Rathke, Martin Heinrich. Untersuchung über die bernstein-insecten. (Oken's Isis, 1829, 413.) 4°. Leipzig. 1829.

Mentions his large collection of amber insects collected in company with Berendt; the Coleoptera had been studied by Heyden and Schmidt, the Ichneumonidae by Gravenhorst, the Diptera by Meigen and Wiedemann.

Reuss, A. E. Geognostische skizzen aus Böhmen. 1er theil: Die umgebungen von Teplitz und Bilin in beziehung auf ihre geognostischen verhältnisse; ein beitrag zur physiographie des böhmischen mittelgebirges; mit 1 karte und 9 tafeln. 8°. Leitmeritz. 1840.

Reuss, A. E.—Continued.

Not seen; according to Deichmüller this work contains a reference on p. 143 to the occurrence of fossil insects at Bilin.

—— Die geognostischen verhältnisse des Egerer bezirkes und des Ascher gebietes in Böhmen. (Abhandl. k. k. geol. reichsanst., 1: 1–72, map.) 4°. Wien. 1852.

Mentions, p. 58, the occurrence of impressions of Coleoptera and Diptera, generally very indistinct, in the cyprismergel of Kuttensee. An exception is noted in a single Dipteron, which is figured and which Kollar places near Penthetria.

Richter, Georg Gottfried. Gazophylacium mineralium oder Mineralienkabinet. 8°. Leipzig. 1719.

Not seen; said to contain something on Oeningen insects.

Rolle, Friedrich. Ueber ein vorkommen fossiler pflanzen zu Obererlenbach (Wetterau). (Neues jahrb. f. mineral., 1877, 769–784.) 8°. Stuttgart. 1877.

Merely mentions, pp. 772–773, the occurrence of remains of insects in the pliocene? beds of Ober Erlenbach.

Rondani, Camillo. Lettre sur les insectes du succin. (Rev. zool., 3: 366–370.) 8°. Paris. 1840.*

Followed, p. 370, by remarks by Guérin-Méneville. Rondani makes some corrections of generic determinations in Guérin's supplement to Maravigna's paper. Guérin speaks only of the imperfection of the specimens. See also Maravigna, C.

Roy, C. W. van. Ansichten über entstehung und vorkommen des bernsteins, so wie praktische mittheilungen über den werth und die behandlung desselben als handelswaare. 8°. Danzig. 1840.

Not seen.

Samouelle, George. See **Murchison, R. I.**

Saporta, *Marquis* Gaston de. Examen des flores tertiaires de Provence. (Heer, Climat pays tert., pp. 133–171.) 4°. Winterthur. 1861.

At the end of the section on the flora of Aix he refers, pp. 152–153, to the insects of the gypsum beds and their relations to the vegetation of the epoch.

—— Études sur la végétation du sud est de la France à l'époque tertiaire. Suppl. I. Révision de la flore des gypses d'Aix; fasc. I: généralités. 8°. Paris. 1872. pp. 79, pl. 2. (Ann. sc. nat., (5), bot., 15: 277–351, pl. 15–16.) 8°. Paris. 1872.

Saporta, G. de—Continued.

Notices the insects of Aix on pp. 70–71 [342–343]. Gives also a note by A. Giard, p. 69 [341], suggesting the presence at Aix of certain plants, from the occurrence of beetles presumed to feed upon them.

—— See also **Heer, O.**

Scarabelli, Giuseppe. See **Massalongo, A. B. P., e Scarabelli, G.**

Schau-platz der natur oder gespräche von der beschaffenheit und den absichten der natürlichen dinge, etc. Dritter theil. 8°. Wien und Nürnberg. 1748. pp. (22), 592, (11), front., pl. (33).

Refers to the occurrence of flies and beetles in amber, p. 347. A figure of a fossil odonate larva also appears on the plate opposite p. 416 (fig. F).

Scherper. See **Schöberlin, E.**

Scheuchzer, J. J. Beschreibung der natur-geschichten des Schweizerlands dritter theil enthaltende vornemlich eine oder die höchsten alpgebirge an. 1705 getahne reise. 4°. Zürich. 1708. pp. (4), 208, pl. (9).

Refers, p. 68, to the occurrence of flies and spiders in amber (agdstein).

—— Museum diluvianum quod possidet J. J. S. 16°. Tiguri. 1716. pp. (12), 107, (4), front.

Records four fossil insects, p. 106, from Oeningen, Monte Bolca, and Querfurt.

—— Physique sacrée, ou Histoire naturelle de la bible. Traduite du latin; enrichie de figures en taille douce, gravées par les soins de Jean-André Pfeffel. 8 vol. f°. Amsterdam. 1732–1737.

Vol. I, tab. 53, figs. 23–25, p. 68, gives figures of a beetle and odonate larva from Oeningen, and an odonate from Verona, which Heer afterwards determines. Original edition not seen.

Schilling, Peter Samuel. See **Gravenhorst, J. L. K.**

Schlechtendal, D. von. Mittheilungen über die in der sammlung aufbewahrten originale zu Germar's "Insekten in bernstein eingeschlossen" mit rücksicht auf Giebel's "Fauna der vorwelt." (Zeitschr. ges. naturw., 61: 473–491.) 8°. Berlin. 1888.

Subjects nearly all the originals of Germar's paper to a new and careful study, with one or two new illustrations.

Schmidt, Wilhelm Ludwig Ewald. See **Rathke, M. H.**

Schöberlin, Edmund. Der Oeningen stinkschiefer und seine insektenreste. (Soc. entom., 3: 42, 51, 61, 68–69.) 4°. Zürich. 1888.

A general account of the insects found at Oeningen, based mainly on Heer's researches. Nearly eleven hundred species are now known and several genera are here for the first time recorded, the result of studies by Lindner, Massmann, and Scherper.

Schweigger, August Friedrich. Beobachtungen auf naturhistorischen reisen ; anatomisch-physiologische untersuchungen über corallen ; nebst einem anhange, bemerkungen über den bernstein enthaltend. 4°. Berlin. 1819. pp. 8, 128, (4), pl. 8, tab. 12.

The Bemerkungen über den bernstein occupy pp. 101–127 and pl. 8, and contain in foot-notes extended descriptions of a few insects, figured carefully on the plate, but part at least of which have since been recognized as royal insects.

Scudder, S. H. Results of an examination of a small collection of fossil insects obtained by Prof. William Denton in the tertiary beds of Green River, Colorado. (Proc. Bost. soc. nat. hist., 11 : 117–118.) 8°. Boston. 1867.

The same : with slight additions and without title. (Hollister, Mines of Colorado, pp. 360–363.) 12°. Springfield. 1867.

A bare statement of the relations of the insects. An abstract occurs in Amer. nat., 1 : 56. 8°. Salem. 1867.

———— [Fossil insects found at the petrified fish-cut, Green River.] (Hayden, Sun pict. Rocky mount. scen.,p. 95.) 4°. New York. 1870.

Mentions in general terms the affinities of an ant and a couple of flies.

———— Fossil insects from the Rocky Mountains. 8°. Salem. 1872. pp. 4. (Amer. nat.,6 : 665–668.) 8°. Salem. 1872.

A general notice of a collection of insects made by Richardson in the tertiary Green River deposits of Wyoming. Noticed in Trans. ent. soc. Lond., (2), 1874, 47–48. 8°. London. 1874.

———— The insects of the tertiary beds at Quesnel (British Columbia). 8°. [Montreal. 1877.] pp. 15. (Rep. progr. geol. surv. Can., 1875–'76, 266–280.) 8°. [Montreal.] 1877.

Describes twenty-four species of various orders.

Scudder, S. H.—Continued.

TRANSLATION : Les insectes des lits tertiaires de Quesnel. (Rapp. opér. expl. géol. Can., 1875–'76, 294–310.) 8°. [Montreal.] 1877.

———— The first discovered traces of fossil insects in the American tertiaries. (Bull. U. S. geol. geogr. surv. terr., 3 : 741–762.) 8°. Washington. 1877.

Describes the insects obtained by Denton in the White River beds, Colorado. Thirty-three species, mostly Diptera, are described and others enumerated.

———— Additions to the insect fauna of the tertiary beds at Quesnel (British Columbia). 8°. [Montreal. 1878.] pp. 8. (Rep. progr. geol. surv. Can., 1876–'77, 457–461.) 8°. [Montreal.] 1878.

Describes six more species of different groups.

TRANSLATION : Additions à la faune entomologique des lits tertiaires de Quesnel, Colombie britannique. (Rapp. opér. comm. géol. Can., 1876–'77, 514–522.) 8°. [Montreal.] 1878.

———— An account of some insects of unusual interest from the tertiary rocks of Colorado and Wyoming. (Bull. U. S. geol. geogr. surv. terr., 4 : 519–543.) 8°. Washington. 1878.

Describes ten insects of different orders, among them a remarkably perfect butterfly, Prodryas persephone, and eggs and egg clusters of a gigantic sialid, Corydalites fecundum, the last from the Laramie beds.

———— The fossil insects of the Green River shales. (Bull. U. S. geol. geogr. surv. terr., 4 : 747–776.) 8°. Washington. 1878.

Describes fifty-five species of different groups with notes on seventeen others.

———— The fossil insects collected in 1877, by Mr. G. M. Dawson, in the interior of British Columbia. (Rep. progr. geol. surv. Can., 1877–'78, B : 176–186.) 8°. Montreal. 1879.

Describes sixteen species of different orders. Also published separately with half-title on cover: Insects from the tertiary beds of the Nicola and Similkameen rivers, British Columbia. 8°. [Montreal. 1879.] pp. 11.

TRANSLATION : Insectes fossiles recueillis en 1877, par M. G. M. Dawson, dans l'intérieur de la Colombie-britannique. (Rapp. opér. comm. géol. Can., 1877–'78, B : 211–223.) 8°. [Montreal. 1879.]

Scudder, S. H.—Continued.

—— The insect basin of Florissant. (Psyche, 3: 77.) 8°. Cambridge. 1880.

Exhibition of plates of fossils.

—— The tertiary lake basin at Florissant, Colorado, between South and Hayden Parks. (Bull. U. S. geol. geogr. surv. terr., 6: 279–300, map.) 8°. Washington. 1881.

The first half is descriptive of the locality and its geology; the paleontological portion is mainly devoted to insects and plants, of which a running systematic review is given. The conclusion is reached that the beds, the most prolific of insects in the world, "belong in or near the oligocene." An abstract will be found in Harv. univ. bull., 2: 267. 4°. Cambridge. 1881. It was also read before the Natural History Society of Boston. (Proc. Bost. soc. nat. hist., 21: 81.) 8°. Boston. 1881.

EXTRACT: Insects of the amyzon shales of Colorado. (Amer. nat., 16: 159–160.) 8°. Philadelphia. 1882.

Quotes some of the general results obtained.

REPRINT: with same title as original. (Twelfth Rep. U. S. geol. geogr. surv. terr., 1878: 271–293, map.) 8°. Washington. 1883.

Contains considerable additions, especially in the Arachnida and Neuroptera, where comparisons are instituted with European and other American fossils.
It is also reprinted with slight additions and omissions in the introduction to the author's Tertiary Insects of North America, 1890.

—— Administrative report U. S. geological survey for the year 1885–1886. (Ann. rep. U. S. geol. surv., 7: 127.) 8°. Washington. 1888.

Progress of work on tertiary Coleoptera and Diptera.
See also Section VIII.

—— The fossil insect localities in the Rocky Mountain region. (Psyche, 5: 362.) 4°. Cambridge. 1890.

The relative abundance of individuals in the various orders of insects and especially in the Hymenoptera and Coleoptera is very different at Florissant from what it is at the other localities, while the latter are similar to one another.

—— The tertiary insects of North America. pp. 734, pl. 28. 4°. Washington. 1890. (Rep. U. S. geol. geogr. surv. terr., xiii.)

Forms vol. 2. of the Fossil Insects of North America.
Treats in full the Myriapoda, Arachnida, and lower orders of Hexapoda from all American

Scudder, S. H.—Continued.

tertiary deposits, while in the Coleoptera, Diptera, and Hymenoptera, only the species found elsewhere than at Florissant are described. It includes of Myriapoda 1 sp., Arachnida 34, Neuroptera 66, Orthoptera 30, Hemiptera 264, Coleoptera 112, Diptera 79, Lepidoptera 1, and Hymenoptera 23, in all 612 species. All are displayed in a tabular view at the end. A few notes occur on European species.

—— See also Peale, A. C.

Selwyn, Alfred R. C., and **Dawson, George Mercer.** Descriptive sketch of the physical geography and geology of the Dominion of Canada. 8°. Montreal. 1884. 55 pp.

Copies, p. 55, the statement in Dawson's paper, q. v.

Sendel, Nathaniel. De succino indico, ad virum nobilissimum atque experientissimum dominum Johannem Philippum Breynium epistola. prodromi loco electrologiae sunc propediem edendae scripta. (Breyn, Melon. petref. mont. Carmel, 35–48.) 4°. Lipsiae. 1722.

Entitled on Breyn's title-page: De pseudo-succino, quod paucos ante annos ex Africa in Belgium deferri corpit.' It evidently refers to copal, and mentions, p. 40, the occurrence of insects "nostris similin." Noticed in Bibl. germ., 5: 121.

—— Verschiedene erinnerungen von dem succino prussico. (Contin. gelehrt. Preussen, 1725, quart. 2, no. 3.) 8°. Thorn. 1725.

Not seen. Title furnished by Dr. Hagen.

—— Nathanaelis Scudelii . . . electrologiae per varia tentamina historica ac physica continuandae missus primus De perfectione succinorum operibus naturae et artis promota testimoniisque rationis et experientiae demonstrata. 4°. Elbingae 1725. pp. 56.

Reviewed in Acta erud., 1725, 374–376. 4°. Lipsiae. 1725.

The same. Missus secundus, De mollitie succinorum et inde emergentibus contentis variis animalibus, vegetabilibus, mineralibus atque aquosis. 4°. Elbingae. 1726. pp. 64.

The same. Missus tertius, De prosapia succinorum et eorum variis affectionibus, vi electrica, colore, odore, sapore. 4°. Elbingae. [1722.] pp. 56.

These titles are furnished by Dr. Hagen, who also quotes, but unverified by him, the English translation in Acta germ., 1743, 340–353, 360–368, 389–405. 4°. London. 1743.

Sendel, N.—Continued.

—— Historia succinorum corpora aliena involventium et naturae opere pictorum et caelatorum ex regiis Augustorum cimeliis Dresdae conditis aeri insculptorum conscripta. 1°. Lipsiae. 1742. pp. 10, 328, tab. 13.

A large part of the book and nearly seven of the plates are given up to insects, but amber and copal insects are, as is well known, not distinguished, and the book has therefore far less value and interest than it otherwise would possess.

Reviewed in Nova acta erud., 1743, 49-56. 4°. Leipzig. 1743 (the insects mentioned on pp. 50-52); also, according to Boehmer (references unverified) in Nouv. bibl. germ., 3: 1; 6: 385; Commerc. litt. med. sc. nat., 23: 177. 4°. Norimbergiae. 1742;—Leipz. gel. zeit., 1742, 414. 8°. Leipzig. 1742;—Zuverl. nachr. gegenw. zust. wiss., 34: 778. 8°. Leipzig.

See also Guérin-Méneville, F. E., in Section I.

Serres, Pierre Marcel Toussaint de. Suite du Mémoire sur les terrains d'eau douce, ainsi que sur les animaux et les plantes qui vivent alternativement dans les eaux douces et dans les eaux salées. (Journ. phys., 87: 161-178.) 4°. Paris. 1818.

Refers, p. 173, to impressions of insects, principally apterous and among them imlada, from the tertiaries of Castelnaud, France.

—— Note sur les arachnides et les insectes fossiles, et spécialement sur ceux des terrains d'eau douce. (Ann. sc. nat., 15: 98-108.) 8°. Paris. 1828. (Ferr., Bull. sc. nat., 15: 181-189.) 8°. Paris. 1828.

An extract from the next work, pp. 207-233, published in advance.

TRANSLATION: Bemerkung über die fossilen arachniden und insecten, besonders über diejenigen, welche in der süsswasserformation vorkommen. (Thon, Entom. archiv., 2, ii: 74-77.) 4°. Jena. 1829.

ABSTRACT: Notiz über fossile arachniden und insecten, und besonders über diejenigen, welche in den niederschlägen des süssen wassers gefunden werden. (Froriep, Notiz. geb. nat.- nhuilk., 22: 337-342.) 4°. Erfurt. 1828.

—— Geognosie des terrains tertiaires, ou Tableau des principaux animaux invertébrés des terrains marins tertiaires, du midi de la France, etc. 8°. Montpellier et Paris. 1829. pp. 92, 277, tableaux 3, pl. 6.

Serres, P. M. T. de—Continued.

Livre 4: Des arachnides et insectes fossiles, et spécialement de ceux des terrains d'eau douce du bassin tertiaire d'Aix, occupies pp. 206-258, and includes a list of nearly 80 genera of Aix insects, besides, pp. 254-258, a Tableau général des arachnides et des insectes fossiles, d'après l'ordre de formations géologiques, in which 105 genera are specified and 226 species enumerated.

—— Notes géologiques sur la Provence. (Actes. soc. linn. Bord., 13: 1-82.) 8°. Bordeaux. 1843.

Contains a list of insects of Aix, pp. 34-44.

—— Note additionnelle à la Notice géologique sur la Provence. (Actes soc. linn. Bord., 13: 83-91.) 8°. Bordeaux. 1843.

Continuation of preceding, but with nothing on insects.

Shuckard, William Ed. See Burmeister, H.

Sivers, Henricus Jacobus. Cyriosorvm niemdorpiensivm. 4°. Lybecae. 1732.

Specimen IV, sistens svccinorvm descriptionem occupies pp. 65-81, with a plate. Reference furnished by Dr. Hagen.

Smith, E. J. A'Court. Discovery of remains of plants and insects. (Nature, 11: 88.) 4°. London. 1874.

Notice of fossil insects in the tertiary beds at Garnet Bay, Isle of Wight.

Smith, Frederick. See Woodward, H.; and Zaddach, E. G.

Sordelli, F. Note sopra alcuni insetti fossili di Lombardia. 8°. Firenze. 1882. 12 pp. (Bull. soc. entom. ital., 14: 221-235.) 8°. Firenze. 1882.

Scattered notes and descriptions of insects from late tertiary, particularly pleistocene deposits.

Stainton, Henry Tibbats. See Bolton, J.

Stein, Johann Philipp Emil Friedrich. Drei merkwürdige bernstein-insecten. (Mitth. münch. entom. ver., 1: 28-30.) 8°. München. 1877.

Describes two Coleoptera and one Hymenopteron (Myrmar).

Steinbeck, A. Ueber die bernsteingewinnung bei Brandenburg an der Havel. 12°. Brandenburg. 1841. (Neue not. natur- heilk., 14: 257-263.) 4°. Weimar. 1840.

Separate publication not seen. Notices, p. 262, collections of amber insects made by Schirrmeister and himself, showing the fauna to be the same as that of the amber of the Baltic coast.

Steinbeck, A.—Continued.

An abstract will be found in Neues Jahrb. mineral., 1844, 121–122. 8°, Stuttgart, 1844.

Sternberg, K. Insektengänge im blatte der Flabellaria borassifolia. (Verhandl. gesellsch. vaterl. mus. Böhm., 1836: 34–35, pl. 1, figs. 3–4.) 16°. Prag. 1836.

Showing mines of an insect "ganz so, wie es die larven der blattschaben in dem parenchym der blätter jetztweltlicher pflanzen hinterlassen."

Stizenberger, Ernst. Uebersicht der versteinerungen des grossherzogthums Baden. 8°. Freiburg i. B. 1851. pp. 144.

A list of Oeningen insects compiled from the first two parts of Heer's work occurs on pp. 95–101 with references to collections: and on p. 119 a reference to eggs and larvæ of beetles and flies from the alluvium of the Rhine.

Tournal. [Tertiary of Armissan.]

Tournal is said to have published some reference to fossil insects in a work on the above subject. I have been unable to verify it.

Troost, Gerard. Description of a variety of amber, and of a fossil substance supposed to be the nest of an insect discovered at Cape Sable, Magothy River, Ann-Arundel County, Maryland. (Amer. journ. sc. arts, 3: 8–15.) 8°. New Haven. 1821.

Considers the nest found in a stratum of lignite, and which is described on pp. 10–11, to be "a kind of comb or nidus made by some insects around the twigs . . . of a tree."

TRANSLATION: Bernstein mit gall-insekten nestern vorkommend in Maryland in Nord-Amerika nach dem Dr. Troost zu Baltimore, frei bearbeitet von Gilbert. (Ann. phys., 70: 297–303.) 8°. Leipzig. 1822.

Accompanied by a foot-note by Gilbert, and followed, pp. 303–304, by a Zusatz zu diesem aufsatze by the same, in which certain similar appearances in European amber are noted. Gilbert considers them galls.

TRANSLATION: Ueber das vorkommen des bernsteins zu Cap-Sable in Nord-Amerika. (Jahrb. chem. phys., 4 [Journ. chem. phys., 34]: 434–439.) 16°. Nürnberg. 1822.

The final notes are omitted and the phraseology slightly altered.

ABSTRACT: Vorkommen des bernsteins in Nord-Amerika. (Arch. bergb. hüttenw., 6: 416.) 16°. Berlin. 1822.

Makes no reference to the "insect nests."

TRANSLATION: Beschreibung einer varietät von bernstein (amber) und eines

Troost, G.—Continued.

fossils, wahrscheinlich des nestes eines insekts, entdeckt bei Cap Sable, am Magothy-fluss, in Ann-Arundel County, Maryland. (Schrift. phys.-ökon. gesellsch. Königsb., 11: 54–58.) 4°. Königsberg. 1871.

The translation, which is by Dr. Hagen, omits only about a page of unimportant matter in the concluding notes. It is embodied in an article by Dr. Berendt, entitled Ueber eine von Dr. G. Troost in Baltimore im jahre 1821 im American-journal of science gegebene beschreibung eines bernstein-vorkommens bei Cap Sable in Maryland; and is preceded and followed by remarks of Dr. Berendt, which, however, do not refer to the insect nests.

—— See also **Ballenstedt, J. G. J.**; and **Hagen, H. A.**

Unger, Franz. Ueber die pflanzen und insekten reste von Radoboj in Kroatien. (Unger, Reise notizen, 1838, 26–33.) 8°. [Wien. 1840?]

Not seen; an abstract will be found in Neues jahrb. miner., 1840, 374–377. 8°. Stuttgart. 1840. Insects are mentioned an p. 377; there are no Coleoptera nor Lepidoptera; Diptera and Hymenoptera are most abundant; Neuroptera, Orthoptera and Hemiptera rarer. A single spider was found. The fauna is tropical rather than European.

The Reise notizen are referred to in no bibliographies. They appeared separately for 1838 and 1839, and were probably extracts from some local paper. Cf. Neues jahrb. mineral., 1840, 726.

—— Chloris protogaea. Beiträge zur flora der vorwelt. f°. Leipzig. 1840–'47. Heft 1, pp. 4, 4, 1–16, pl. 1–5 (1840);—heft 2–3, pp. 5–24, 17–44, pl. 6–15 (1842);—heft 4–5, pp. 45–92, pl. 16–25 (1843);—heft 6–7, pp. 25–110, pl. 26–35 (1845);—heft 8–10, t. p., ded., pp. 93–150, pl. 36–50 (1847).

Insects are figured on pl. 4, 5, 11, 14, 15, 22, 28, 40, 44, 45, 46. All are from Radoboj.

—— See also **Heer, O.**

Vaillant, Léon. Sur un geckotien de l'ambre jaune. (Bull. soc. philom. Paris, 10: 65–67.) 8°. Paris. 1873.

Mentions, p. 67, a couple of insects accompanying the reptile, afterwards more fully treated by Oustalet.

Valentini, Michael Bernhard. Museum museorum oder Vollständige schau bühne aller materialien und specereyen etc. Zweyte edition. 3 v. f°. Franckfurt am Mayn. 1714. Vol. 1, 3 t. p., pp. (24), 520 (16), 76, (4), 119;—vol. 2, 3 t. p., pp.

Valentini, M. B.—Continued.
(18), 196, (4), 116;—vol. 3, pp. (4), 218,
(12), plts., figs.

Also with the title: D. Valentini schau bühne
oder Natur- und materiallen-kammer, auch ostindi-
anische send-schreiben und rapporten. Contains
various references to insects in amber, original or
quoted; see especially i. p. 516, ii. p. 60, and ii.
anhang, pp. 93, 99.

Vogel, Rudolf Augustin. Practisches
mineral-system. Zweyte vermehrte und
verbesserte ausgabe. 8°. Leipzig. 1776.
pp. (22), 582.

Not seen. The first edition was published in
1762, and is said to contain reference to Oeningen
insects.

Walch, Johann Ernst Immanuel. See
Knorr, G. W.

Walchner, Friedrich August. Darstellung der geologischen verhältnisse des
süsswasser-mergels von Oeningen im ba-
dischen seekreis und seiner fossilen flora
und fauna. 8°. Karlsruhe. 1850.

Separately printed from his Handbuch der geognosie zum gebrauche bei seinen vorlesungen, und
zum selbststudium, mit besonderer berücksichtigung der geognostischen verhältnisse des gross-
herzogthum Baden. 2°. aufl. 8°. Karlsruhe.
1847-1851. (pp. 956 et seq.) Neither seen.

——— Darstellung der geologischen
verhältnisse des mainzer tertiärbeckens
und seiner fossilen fauna und flora. 8°.
Carlsruhe. 1850. pp. 75.

Contains, p. 57, a list of insecta referred to
twelve genera, only two of the species receiving
names,—Phryganea mombachiana and P. blumi.
Separately printed from his Handbuch der
geognosie, etc., as above. 2°. aufl. 8°. Karlsruhe. 1847-51.

Walker, John Francis. Fossil insects
in the Bournemouth leaf beds. (Geol.
mag., 7: 240.) 8°. London. 1870.

A bibliographical reference.

Wanklyn, A. Description of some
new species of fossil ferns from the Bournemouth leaf-bed. (Ann. mag. nat. hist.,
(4), 3: 10-12, pl. 1.) 8°. London. 1869.

Mentions an undetermined insect.

Westwood, J. O. See **Jones, T. R.**

**Wiedemann, Christian Rudolph Wil-
helm.** See **Rathke, M. H.**

Wigand, Johannes. Vera historia de
succino borvssico. De alce borvssica &
de herbis in Borussia nascentibus. Item
de sale creatvra Dei salvberrima conside-

Wigand, J.—Continued.
ratio methodica & theologica per Iohan-
nem VVigandvm D. Qvondam episcopum
pomozaniensem. Iam vero primvm in
studiosae iuuentutis gratiam in lucem
edita. Studio et opera Iohannis Rosini
pastoris vvickerstadensis. 12°. Ienae.
[MD]XC, ff. (12), 153, (5).

In a section: De vermiculis in succino, ff. 26-29,
he mentions culices, formicae, aranei parul papiliones.

Woodward, H. On the occurrence
of Branchipus (or Chirocephalus) in a
fossil state, associated with Archaeonis-
cus and with numerous insect remains
in the eocene fresh-water limestone of
Gurnet Bay, Isle of Wight. (Geol. mag.,
n. s., 5: 82-89.) 8°. London. 1878.

Abstract of the next, with slightly differing
title, but published earlier than it, and containing a fuller list of insects.

——— On the occurrence of Branchipus (or Chirocephalus) in a fossil state,
associated with Eosphaeroma and with
numerous insect remains in the eocene
fresh-water (Bembridge) limestone of
Gurnet Bay, Isle of Wight. (Quart.
journ. geol. soc. Lond., 35: 342-350, pl.
14.) 8°. London. 1879.

The insects are mentioned on p. 344, mostly in
a List of insect remains from Gurnet Bay, near
Cowes, Isle of Wight, determined by the late
Frederick Smith. One hundred and twenty-five
specimens are mentioned but only eighteen genera
or families are specified and one species.

Zaddach, Ernst Gustav. Ueber die
bernstein- und braunkohlenlager des Sam-
landes; erste abhandlung. (Schrift.
phys.-ökon. gesellsch. Königsb., 1: 1-44,
pl. 1-4.) 4°. Königsberg. 1860.

Notices, pp. 3-4, the numbers of insects found
in amber, all distinct from living forms; and on
pp. 20-21 gives a table of the number of genera
and species of the different orders of insects,
with special mention of the remarkable genera
Archaea and Amphientomum.

——— Amber; its origin and history,
as illustrated by the geology of Samland.
(Quart. journ. science, 5: 167-185, pl. (2).)
8°. London 1868.

Mainly compiled from the author's previous
writings on the amber beds. It contains, however, as new matter, a plate of amber insects
with explanation by Frederick Smith, and on pp.
184-185, a list of the principal works on amber
and the organic remains preserved in it, furnished
by the editors.

.˙. Ami **Boué** is said to have been the first (Journ. geol., 3: 105) to have referred to the insects of Radoboj, but I have been unable to verify the reference; and John **Ray** is stated to have made some references to fossil insects on pp. 78 and 92 of his Historia insectorum; but I have examined the work for such references unsuccessfully.

VII.—SPECIAL FOR CENOZOIC TIME.

VIIa.—Cenozoic Myriapoda.

.˙. See also under Section I and Section VI.

Bertkau, P. Einige spinnen und ein myriapode aus der braunkohle von Rott. (Verhandl. naturh. verein preuss. Rheinl. u. Westf., (4), 5: 346–360, taf. 5.) 8°. Bonn. 1878.

Describes and figures Julus antiquus Heyd. MSS. See same title in Section VIIb.

Cotta, Bernhard. Ueber Julus terrestris, als jugendliche versteinerung. (Neues jahrb. f. miner., 1833, 392–394, pl. 5.) 8°. Stuttgart. 1833.

Description and figure of this species as found fossil near Dresden in kalksintergänge in gneiss. Münster, loc. cit., p. 68, speaks of these as lituiten-artige röhre.

Heyden, C. von. See **Bertkau**, P.

Münster, G. See **Cotta**, B.

VII b.—Cenozoic Arachnida.

.˙. See also under Section I and Section VI.

Bertkau, P. Einige spinnen und ein myriapode aus der braunkohle von Rott. (Verhandl. naturh. verein preuss. Rheinl. u. Westf., (4), 5: 346–360, taf. 5.) 8°. Bonn. 1878.

A careful description of eight species, of which six are new; all are figured. See same title in Section VIIa.

———— Ueber einige fossile arthropodenreste aus der braunkohle von Rott. (Sitzungsb. niederrhein. gesellsch. Bonn, 1878, 70–71.) 8°. Bonn. 1878.

Principally concerned with the arachnids described in his previous paper, and especially with Argyroneta antiqua.

Breyn, Johann Philipp. Observatio de succinea globo, plantae ejusdam folio impregnata, rarissima. (Phil. trans., 34; 154–156, pl., fig. 2.) 4°. London. 1728.

Mentions a spider in amber. See same title in Section VIIg.

Brongniart, C. J. E. Note sur une Aranéide fossile des terrains tertiaires d'Aix (Provence). (Ann. soc. ent. France, (5), 7: 221–224, pl. 7, i.) 8°. Paris. 1877.

Describes Attoides eresiformis. Noticed by Dr. Hector George in the Feuilleton of La constitutionnel for 21 Nov., 1877. See also Girard, M.

George, H. See **Brongniart**, C. J. E.

G[irard], M. Une très-ancienne araignée. (La nature, 6: 144, figs. 1–4.) 4°. Paris. 1878.

A popular account of Attoides eresiformis described by Brongniart.

Gourret, Paul. Recherches sur les arachnides tertiaires d'Aix en Provence. (Rec. zool. suisse, 4: 431–496, pl. 20–23.) 8°. Geneva. 1887.

Describes twenty-three species, almost all referred to new genera. Besides the Araneides, there are two Acari, two Opiliones, and one pedipalp. A discussion of their faunal relations at the close concludes them to have close affinities with Mediterranean forms.

Heyden, C. von. Nachricht von fossilen gallen auf blättern aus den braunkohlengruben von Salzhausen. (Ber. oberhess. gesellsch. nat.- heilk., 8: 63.) 8°. Giessen. 1860.

Probably the gall of a Phytoptus, on Salix.

Karsch, F. Neue milben in bernstein. (Berl. entom. zeitschr., 28: 175–176, figs.) 8°. Berlin. 1884.

Three species of Nothrus described.

McCook, H. C. A new fossil spider, Eoatypus woodwardii. (Proc. acad. nat. sc. Philad., 1888, 200–202, 2 figs.) 8°. Philadelphia. 1888.

One of the Atypinae from the Isle of Wight.

Menge, A. Ueber die scheerenspinnen, Chernetidae. 4°. [Danzig. 1855.] pp. 43, pl. 5. (Neueste schrift. naturf. gesellsch. Danzig, 15, heft 2. [art. 2].) 4°. Danzig. 1855.

Seven of the fifteen species described and figured are from amber.

Menge, A.—Continued.

—— Ueber einen scorpion und zwei spinnen im bernstein. 8°. Danzig. 1869. pp. 9, 3 figs. (Schrift. naturf. gesellsch. Danzig, (2), 2, no. 10.) 8°. Danzig. 1869.

Detailed description and figures of Tityus cognatus, Clostes priscus, and Gerdia myura.

Scudder, S. H. Fossil spiders. (Harv. univ. bull., 2 : 302–303.) 8°. Cambridge. 1882.

Reviews the arachnid fauna of Florissant, Col., and shows its relation to the forms of the European tertiaries.

REPRINT: Our knowledge of fossil spiders. (Field natur., 1 : 61–63.) sm. 4°. Manchester (Engl.), 1882.

Simon, Eugène. Description d'un genre nouveau d'Arachnides et remarques sur la famille des Archæidæ. 8°. Genova. [1884.] pp. 6, figs. 1–5. (Ann. mus. civ. stor. nat., 20 : 182–187, figs. 1–5.) 8°. Genova. [1884.]

Erlanchemna of Madagascar and Landona of Kongo are placed with Archæa from amber in the family Archæidæ.

—— Note complémentaire sur la famille des Archæidæ. 8°. Genova. 1884. pp. 8, figs. 1–7. (Ann. mus. civ. stor. nat., 20 : 373–380, figs. 1–7.) 8°. Genova. 1884.

Describes and figures Archæa pougnetti from amber.

Turpin, Pierre Jean François. Note sur le terrain qui contient le tripoli de Bilin, en Bohême, par M. Élie de Beaumont; suivie de l'examen des débris organiques que renferme une des couches de ce terrain, par M. Turpin. (Comptes rendus acad. sc., 7 : 501–503.) 4°. Paris. 1838.

The Note de M. Turpin occupies pp. 502–503. Mention is made, p. 502, of the leg of an insect "très probablement d'un Acarus," as found in the earth.

Walckenaer, C. A. Tableau des aranéïdes, ou caractères essentiels des tribus, genres, familles et races que renferme le genre Aranea de Linné, avec la désignation des espèces comprises dans chacune de ces divisions. 8°. Paris. 1805. pp. 4, 12, 88, tabl. pl. 9.

On p. 25 catalogues without description Attus fossilis from amber.

Bull 69——6

VIIc.—Cenozoic Neuroptera.

.•. See also under Section I and Section VI.

Andrae, K. J. Beiträge zur kenntnisse der fossilen flora Siebenbürgens und des Banates. Mit zwölf tafeln. pp. 1–48, pl. 1–12. (Abhandl. k. k. geol. reichsanst. Wien, bd. 2, abth. 3, no. 4.) 4°. Wien. 1855.

Figures a Chrysopa, pl. 5, fig. 3, 3a, from Thalheim, with mention on p. 26. See same title in Section VIII.

Berendt, G. C. See Pictet de la Rive, F. J., and Hagen, H. A.

Bosc, Louis [Augustin Guillaume]. Note sur une fossile remarquable de la montagne de Saint-Gérand-le-Puy, entre Moulins et Roanne, département de l'Allier, appelé l'indusie tubuleuse. (Journ. d. mines, 17 : 397–400, pl, 7.) 8°. Paris. an xiii.

The first notice of the remarkable caddis-fly cases of Auvergne, forming the beds of so-called indusial limestone.

Bosniaska, S. de. La formazione gessoso-solfifera e il secondo piano mediterraneo in Italia. (Atti. soc. tosc. sc. nat., 2, proc. verb., 90–100.) 8°. Pisa. 1880.

Refers, p. 93, to Libellula doris and other insects as occurring at two horizons.

Brauer, F. Verzeichniss der bis jetzt bekannten neuropteren im sinne Linné's. pp. 90. n. d. (Verhandl. k. k. zool.-bot. gesellsch. Wien., 18 : 359–416, 711–742.) 8°. Wien. 1868.

Includes the fossil genera and species, and contains, p. 738 (80), a list of the fossil Libellulina.

—— See also Hagen, H. A.

Brongniart, A. Sur les terrains qui paraissent avoir été formés sous l'eau douce. (Ann. mus. hist. nat., 15 : 357–405, pl. 23–24.) 4°. Paris. 1810.

Discusses Indusia tubulata on pp. 302–303.

Capellini, G. La formazione gessosa di Castellina marittima e i suoi fossili. (Mem. accad. sc. ist. Bologna, (3), 4 : 525–603, pl. 1–9.) 4°. Bologna. 1873.

On pp. 539 and 557 catalogues Libellula doris Heer (larva) from Limono, etc.

—— Nota sulla geologia toscana. (Rend. sess. accad. sc. ist. Bologna, 1874–'75, 22–24.) 8°. Bologna. 1875.

Refers, p. 24, to the occurrence of larvæ of a Libellula in tertiary beds near Pane e Vino.

Capellini, G.—Continued.

—— Nuove ricerche sul calcare a amphistegina strati a congeria e calcare di leitha dei Monti Livornesi. (Rend. sess. accad. sc. ist. Bologna, 1874-75, 130-135.) 8°. Bologna. 1875.

Refers, p. 133, to the occurrence of the larvæ of Libellula at Limone, etc.

Croizet, l'abbé. Mémoire sur des débris fossiles de l'Auvergne. Analyse. (Bull. soc. géol. France, 4: 22-26.) 8°. Paris. 1833.

Refers briefly, p. 25, to the Indusia tubulata of Auvergne.

—— Quelques observations sur le Puy de Corent. (Ann. acad. Clerm. Ferr., 11: 135-155.) 8°. Clermond Ferrand. 1832.

Not seen; said to refer to the caddis fly cases of the indusial limestone of Auvergne.

—— et Jobert, ainé. Recherches sur les ossemens fossiles du département du Puy-de-Dôme. 4°. Paris. 1828. pp. (8), 224, (2), map, sect. 8, pl. 48, in several series.

Refers, p. 25, to the occurrence of Indusia tubulata in the calcareous marls of the Auvergne tortiarios.

Defrance, J. L. M. Indusio. (Dict. sc. nat., 23: 411-412.) 8°. Paris. 1822.

Notice of the indusial limestone of Auvergne, and the fossil phryganid cases of which it is composed.

Giard, A. Les coléoptères fossiles d'Auvergne par M. Oustalet; remarques critiques. (Bull. scient. dép. Nord, (2), 1: 56-62, 109-118.) 8°. Lille. 1878.

The Neuroptera as well as the Coleoptera are discussed. See same title in Section VIIf.

Goss, H. Notes on a fossil wing of a dragon fly, from the Bornemouth leaf beds. (Entom., 11: 193-195, fig.) 8°. London. 1878.

Referred to Æschna.

Hagen, H. A. Ueber die nonropteren der bernstoin fauna. (Vorhandl. zool.-bot. ver. Wien, 4: 221-232.) 8°. Wien. 1854.

A systematic review of the nearly 900 specimens examined by the author. The Sitzungsberichte of the same volume, pp. 76-78, contain the remarks of Brauer, comparing the results reached by Hagen with those of Loew and Göppert for Diptera and plants; and the comments of von Hauer, who indicates the places where amber is said to occur in older formations, but never with insect or plant remains.

Hagen, H. A.—Continued.

—— Zwei libellen aus der braunkohle von Sieblos. (Palæontogr., 5: 121-124, tab. 24.) 4°. Cassel. 1858.

Description of Heterophlebia jucunda and Lestes vicina.

—— Asealaphus proavus aus der rheinischen braunkohle. (Palaeontogr., 5: 125-126, tab. 25.) 4°. Cassel. 1858.

Detailed description and comparison with living forms.

—— Petalura ? acutipennis aus der braunkohle von Sieblos. (Palaeontogr., 8: 22-26, taf. 3, figs. 1-4.) 4°. Cassel. 1859.

Detailed description, and discussion of its systematic position.

—— Neuroptera aus der braunkohle von Rott in Siebengebirge. (Palaeontogr., 10: 247-269, taf. 43-45.) 4°. Cassel. 1863.

Extended descriptions of ten species, mostly Odonata, preceded by lists of the insects previously described from the Rhenish brown-coal.

—— Phryganidarum synopsis synonymica. 8°. Wien. 1864. pp. 92. (Verh. zool.-bot. gesellsch. Wien, 14: 799-890.) 8°. Wien. 1864.

Includes the fossil species, twenty-eight in number, of which ten belong to Polycentropus.

—— On some aberrant genera of Psocina. (Ent. monthl. mag., 2: 148-152, 170-172.) 8°. London. 1865-'66.

Describes two species from amber and three from copal, besides seven recent species, being all the ocellate species known.

Same title repeated in Section VIII.

—— Psocinorum et embidinorum synopsis. 8°. Wien. 1866. pp. 22. (Verh. zool.-bot. gesellsch. Wien, 16: 201-222.) 8°. Wien. 1866.

Includes the fossil species, 8 Psocina, 1 Embidina. See same title in Section VIII.

—— Hemerobidarum synopsis synonymica. (Stett. entom. zeit., 27: 369-462.) 16°. Stettin. 1866.

Includes the fossil species, fourteen in number.

—— Beiträge zur kenntniss der phryganiden. (Verh. zool.-bot. gesellsch. Wien, 23: 377-452.) 8°. Wien. 1873.

Hoeninghaus's description of Phryganea rœmbachiana is copied on p. 379, and the insect considered as probably belonging to the Phryganidæ proper.

—— On amber Psocina from Prussia. (Psyche, 3: 279.) 4°. Cambridge. 1882.

Concludes from them that before tertiary times a great development of genera and species had occurred.

Hagen, H. A.—Continued.

—— Beiträge zur monographie der psociden. *At first entitled:* Ueber psociden in bernstein. (Stett. ent. zeit., 43: 217-238, 265-300, 524-526, taf. 1-2: 44: 295-332.) 16°. Stettin. 1882-1883.

Contains a section, Ueber psociden im bernstein, which by error (see p. 265) was printed first under that title. This section is also entitled Bernstein psociden and includes all the portion published in 1882, thirteen species and nine genera (two of them new) being described, while pp. 292-300 are occupied with general considerations drawn from their study.

—— Monograph of the Embidina. (Can. ent., 17: 141-155, 171-175, 190-199, 206-229.) 8°. London (Ont.). 1885.

Extra copies form pp. 141-155, 1-8, 1-10, 1-23. See p. 18 [224].

Describes anew, p. 176 (6), Embia antiqua from amber and refers it to Oligotoma, and under the head of Distribution, p. 223 (17), has a few farther words upon it.

—— See also **Pictet de la Rive,** F. J., und **Hagen,** H. A.; also ⊙.

Hauer, Franz, *ritter* von. See **Hagen** H. A.

Heer, O. Ueber vorweltliche dorfliegen. (Mittheil. naturf. gesellsch. Zürich, 1, ii: 52-54.) 8°. Zürich. 1848.

A brief notice of the fossil dragon flies of Oeningen and Radoboj.

Hepp, Philipp. Ueber die bei Dürkheim aufgefundene versteinerte phryganeen gehäuse. (Jahresb. Pollich., 2: 19-23.) 8°. Neustadt a. d. Haardt. 1844.

An abstract is given in Oken's Isis for 1846, p. 70.

Heyden, C. von. See **Stoehr,** E.

Hoeninghaus, Friedrich Wilhelm. Phryganea monthachiana. 4°. Crefeld. 1811. 1 engr. p. with illustr.

Text in German, accompanied by a French translation of the text (with no heading), 1 p. 8°. See also Hagen, H. A., and Michelin, H.

Jobert, aine. See **Croizet** et **Jobert.**

König, Charles. Icones fossilium sectiles. Centuria prima. 8°. (London). 1825.) no t. p., pp. (4), pl. 19.

Figures on pl. 2 a dragon-fly larva from Oeningen under the name Libellula oeningensis. Two fascicles of this work appear to have been issued: one containing the text and eight plates, the other with eleven plates and no text, neither with title-page. The data given above are taken from de Koninck's copy in the Cambridge Museum.

Kolbe, H. J. Das phylogenetische alter der europäischen psocidengruppen. (Jahresb. westf. prov.-ver. wiss., 10: 18-27.) 8°. Münster. 1882.

Suggestive and interesting discussion of the relative geological age of the genera of Psocidae with a classification in which these points are considered.

—— Neue beiträge zur kenntniss der psociden der bernstein fauna. (Stett. ent. zeit., 44: 186-191.) 16°. Stettin. 1883.

Describes one species each of Philotarsus and Elipsocus and gives a list of amber Psocidae by groups.

—— Der entwickelungsgang der psociden im individuum und in der zeit. (Berl. ent. zeitschr., 28: 35-38.) 8°. Berlin. 1884.

Discussion of the different stages of development which the several groups of Psocidae reach, with a consideration of the fossil forms, especially those from amber.

Kolenati, Friedrich August. Ueber phryganiden im bernstein. (Abhandl. böhm. gesellsch. wissensch., (5), 6: 15.) 4°. Prag. 1851.

Eight species or varieties are named but not described.

Lecoq, Henri. Les époques géologiques de l'Auvergne. 8°. 5 vol. Paris. 1867.

Not seen; according to Oustalet, he discusses Indusia in vol. 2, pp. 335 and 374.

McLachlan, R. [Indusial limestone exhibited.] (Proc. ent. soc. Lond., 1882: 18-19.) 8°. London. 1882.

The fossil larval-cases of which it is composed are referred to Limnophilidae.

Mantell, G. A. A tabular arrangement of the organic remains of the county of Sussex. (Trans. geol. soc. Lond., (2), 3: 201-216.) 4°. London. 1829.

Reference is made on p. 201 to the occurrence of larval cases of Phryganidae in the silt or blue clay of Lewes Levels.

Marion, Antoine Fortuné. See **Saporta,** G. de.

Massalongo, A. B. P. Sopra due larve fossili di Libellula dei terreni mioceni di Sinigallia. (Studii paleont., pp. 22-23, tab. 1, figs. 8-13.) 8°. Verona. 1856.

The larvae are referred to two of Heer's species from Oeningen.

Menge, A. See **Pictet de la Rive,** F. J., und **Hagen,** H. A.

Meyer, C. E. H. von. Indusia. (Ersch u. Gruber, Allg. encycl. wissensch. u. künste, sect. 2, th. 18, s. 136.) 4°. Leipzig. 1840.

Description of the caddis-fly cases of Auvergne.

Michelin, H. [Sur] un travail imprimé de M. Hoeninghaus [sic], relatif à une espèce fossile du genre Phrygane. (Ann. soc. ent. France, (2), 3, bull. ent., 30–31.) 8°. Paris. 1845.

Mere mention of Phryganea grandis. See Hoeninghaus, F. W.

Ouchakoff, Nicolas. Notice sur un termes fossile. (Bull. soc. imp. nat. Mosc., 1838, 37–42, pl. 1.) 8°. Moscou. 1838. (Ann. sc. nat., (2), 13: 204–207, pl. 1 B.) 8°. Paris. 1840.

Description and general remarks.

TRANSLATION: Notice of a fossil termes. (Calc. journ. nat. hist., 2: 74–78.) 8°. Calcutta. 1842.

ABSTRACT: Notiz über eines fossilen termiten. (Neues jahrb. mineral., 1839, 122–123.) 8°. Stuttgart. 1839.

Oustalet, E. See **Giard,** A.

Pictet de la Rive, F. J. Résultat de ses recherches sur les insectes fossiles de l'ordre des névroptères contenus dans l'ambre. (Actes soc. helv. sc. nat., 30: 69–70.) 8°. Genève. 1845.

All specifically distinct from existing forms, but with one exception belonging to existing genera.

——— und **Hagen,** Hermann August. Die im bernstein befindlichen neuropteren der vorwelt. f°. Berlin. [1856.] t. p., pp. 41–125, pl. 5–8. (Berendt, Berust. befindl. org. reste vorw., 2er bd., 1e abth., pars.)

More than seventy species are described in full detail and admirably figured. Under the genera also Hagen gives good accounts of the literature of fossil species and prefaces the whole with general observations on the amber Neuroptera. The work is mostly Hagen's. Some descriptions are by Menge. Description of the larva of a Phasma is also appended, p. 122.

Planchon, Gustave. Étude des tufs de Montpellier au point de vue géologique et paléontologique. 4°. [Montpellier.] 1864.

Not seen; according to Oustalet mention is made of Indusia.

Rouville, Paul de. Géologie des environs de Montpellier. 1855.

Not seen; gives, according to Oustalet, some notice of the remains in the indusial limestone.

Saporta, G. de. Les organismes problématiques des anciennes mers. 4°. Paris. 1884. pp. 6,102, pl. 13.

Larval tubes of phryganids adhering to the leaves of a fossil Nymphaea described, pp. 24–26, and figured, figs. 3–4, by A. F. Marion.

Scheuchzer, J. J. Herbarium diluvianum. f°. Tiguri. 1709. t. p., pp. 44, pl. 10.

Pl. 5, figs. 1–2, p. 16, gives figures and description of an odonate larva from Oeningen and a winged odonate from Monte Bolca.

———. Herbarium diluvianum: editio novissima duplo auctior. f°. Lugduni Batavorum. 1723. 2 t. p., dedic., pp. 119, (5), portr., pl. 14.

The same is found on the same plate, p. 21.

Scrope, George Poulett. The geology and extinct volcanoes of central France. 2d ed. 8°. London. 1858. pp. 17, 258. pl. 17.

Discusses, pp. 10–13, the indusial limestones of France, which he records at the following localities: hills of Gergovia above Romagnat, at the Puys Giron, de Jussat, de la Serre, de Mouton, de Dallet, at Mont Chagny, Mont Jughat, and les Côtes near Clermont; at Davayat near Riom; at Aigneperse, Gannat, Mayet d'école, St. Gerard le Puy, between Jaligny and la Palisse, at Mont Barraud, etc. First edition not seen.

Scudder, S. H. White ants in the American tertiaries. (Harv. univ. bull., 2: 219.) 4°. Cambridge. 1881.

Note on the relation of the six species from Florissant to those of other tertiary deposits: they indicate a warm climate.

——— Notes on some of the tertiary Neuroptera of Florissant, Colo., and Green River, Wyoming Terr. (Proc. Bost. soc. nat. hist., 21: 407–409.) 8°. Boston. 1882.

Summary of general results reached by a study of the American forms.

——— Description of an articulate of doubtful relationship from the tertiary beds of Florissant, Colorado. 4°. [Washington. 1885.] 6 pp., figs. (Mem. nat. acad. sc., 3: 85–90.) 4°. Washington. 1885.

Extended description of Planocephalus, a headless insect presumed to belong to Thysanura, and to form a distinct group therein.

Sismonda, Eugenio. Matériaux pour servir à la paléontologie du terrain tertiaire du Piémont. (Mem. accad. sc. Torino, (2), 22: 391–471, pl. 1–33.) 4°. Torino. 1865.

Sismonda, E.—Continued.

Refers, p. 470, to the occurrence of the larva of Libellula doris in the upper miocene beds of Guarène; it is figured, pl. 17, fig. 6.

Stoehr, Emilio. Notizie preliminari su le piante ed insetti fossili della formazione solfifera in Sicilia. (Bull. com. geol. ital., 1875, 284-287.) 8°. Roma. 1875.

The insects from Girgenti are determined by Dr. von Heyden, p. 280, as larvæ of Libellula doris Heer and L. eurynome Heer, the former in great quantities; both are Oeningen species.

Viquesnel, Auguste. Note sur les environs de Vichy, département de l'Allier. (Bull. soc. géol. France, 14: 145-155.) 8°. Paris. 1842.

Refers on p. 149 to the cases of phryganids.

Wilkinson, C. S. [Fossil insects near Vegetable Creek, New England.] (Proc. Linn. soc. N. S. Wales, 8: 392.) 8°. Sydney. 1883.

Larvæ and pupæ of Ephemeridæ in the tin-bearing tertiary deep leads.

⊙. Termiten im Bernsteinwalde. (Neue preuss. prov.-blätt., (3), 1: 61-61.) 16°. Königsberg. 1858.

A popular account, drawn from Hagen's Monographie der termiten.

VIId.—Cenozoic Orthoptera.

. See also under Section I and Section VI.

Berendt, G. C. Mémoire pour servir à l'histoire des blattes antédiluviennes, traduit de l'allemand par M. Heller. (Ann. soc. ent. France, 5: 539-546, pl. 16.) 8°. Paris. 1836.

Descriptions and figures of the species found in amber

―――― See also Germar, E. F., und Berendt, G. C.

Fritsch, A. Notiz über eine heuschrecke aus der braunkohle von Freudenhain. (Archiv naturw. landesdurchf. Böhmen, bd. 1, sect. 2, p. 276, fig.) 8°. Prag. 1869.

Describes and figures Decticus umbraceus.

Germar, Ernst Friedrich und Berendt, Georg Carl. Die im bernstein befindlichen hemipteren und orthopteren der vorwelt. f°. Berlin. 1856. t.p., pp. 2, 40, pl. 4. (Berendt, Bernst. befindl. org. reste vorw., 2ᵗᵉ bd., 1ᵉ abth.)

Edited, with notes, by Hagen; eight Orthoptera are described and figured, in part by Pictet.

See same title in Section VIIe.

Hagen, H. A. See **Germar, E. F.,** und **Berendt, G. C.**

Heller. See **Berendt, G. C.**

James, J. F. See **Zeiller, R.**

Pictet de la Rive, F. J. See **Germar, E. F.,** und **Berendt, G. C.**

Scudder, S. H. Brief synopsis of the North American earwigs, with an appendix on the fossil species. (Bull. U. S. geol. geogr. surv. terr., 2: 249-260.) 8°. Washington. 1876.

The Note on the fossil species occurs on pp. 259-260, discusses one species already known from the tertiary beds of Florissant, and describes another from the same locality.

―――― Fossil Orthoptera from the Rocky Mountain tertiaries. (Bull. U. S. geol. geogr. surv. terr., 1, ser. 2: 447-449.) 8°. Washington. 1876.

Describes a Labidura and a Homœogamia.

―――― Critical and historical notes on Forficulariæ; including descriptions of new generic forms and an alphabetical synonymic list of the described species. (Proc. Bost. soc. nat. hist., 18: 287-332.) 8°. Boston. 1876. (Scudder, Ent. notes, 5: 27-72.) 8°. Boston. 1876.

Includes the fossil forms, of which eight species are enumerated.

Zeiller, R. Sur des traces d'insectes simulant des empreintes végétales. (Bull. soc. géol. France, (3), 12: 676-680, pl. 30.) 8°. Paris. 1884.

Ridges raised in half dried mud by Gryllotalpa resemble Phymatoderma and Brachyphyllum.

TRANSLATION: On the tracks of insects resembling the impressions of plants. (Journ. Cinc. soc. nat. hist., viii: 49-52.) 8°. Cincinnati. 1885.

Translated by J. F. James.

VIIe.—Cenozoic Hemiptera.

. See also under Section I and Section VI.

Berendt, G. C. Bitte, die bei gräbereien bisweilen vorkommenden fossilen zapfen, fossiles holz und bernstein-insekten betreffend. (Preuss. provinz. blätt., 15: 623-625.) 16°. Königsberg. 1836.

Records, p. 625, the occurrence of a Nepa in amber.

―――― See also Germar, E. F., und Berendt, G. C.

Buckton, G. B. Monograph of the British Aphides. 4 vol. 8°. London. 1875-1883. Vol. 1, t. p., ded., pp. 3, 193, pl. a-o, 1-38;—vol. 2, t. p., pp. 176, pl. 39-86;—vol. 3, t. p., pp. 2, 142, pl. 87-111;—vol. 4, pp. 9, 228, pl. d-i, 115-134.

Vol. 3, pp. 2-4, treat of the successive appearance of insects in time, with special reference to the aphides. Vol. 4 contains a section, pp. 144-178, entitled: Introductory notes on the antiquity of the Hemiptera, and particularly with regard to the Aphidinae as represented in the sedimentary rocks and in amber, in which are described, and figured on plates 131-133, the aphides in amber given by Germar and Berendt, those given by Heer from Oeningen and Radoboj, and (from advance sheets) some of those figured by Scudder from Florissant, the descriptions are based in all cases on the figures of the same, and a few names are given for the first time. He also copies a figure by Millière, and some by Brodie.

Fairmaire, Léon. See Millière, P.

Germar, E. F., und Berendt, G. C. Die im bernstein befindlichen hemipteren und orthopteren der vorwelt. f°. Berlin. 1856. t. p., pp. 2, 40, pl. 4. (Berendt, Bernst. befindl. org. reste vorw., 2er bd., 1e abth.)

Edited, with notes, by Hagen; sixty Hemiptera are described and figured.
See same title in Section VIId.

——— See also Buckton, G. B.

Hagen, H. A. See Germar, E. F., und Berendt, G. C.

Heer, O. Ueber die rhynchoten der tertiärzeit. 8°. Zürich. 1853. pp. 29. (Mitth. naturf. gesellsch. Zürich, 3: 171-197.) 8°. Zürich. 1853.

General account of the relations of the Rhynchota of Oeningen, Radoboj, and Aix to existing faunas, followed by a list of the species described in the third part of his Tertiary Insects. They agree better with the insects of the southern zone than with those of Switzerland, and the Capsini and Riparii characteristic of temperate regions are wholly absent.

——— See also Buckton, G. B.

Heyden, C. von. Fossile insekten aus der braunkohle von Sieblos. Nachtrag. (Palaeontogr., 8: 15-17, pl. 3, figs. 7-9.) 4°. Cassel. 1859.

Description of two Hemiptera and a beetle.
See same title in Section VIIf.

Millière, Pierre. Observations relatives à l'empreinte d'un hémiptère fossile [with note by Signoret, V., and Fairmaire, L.]. (Ann. soc. ent. France, (3), 1: 9-11, pl. 3.) 8°. Paris. 1853.

Millière, P.—Continued.
Under the name of Aphis longicauda, describes an insect from the "schiste marneux" of Ambérieux, Ain.
——— See also Buckton, G. B.

Oustalet, E. Sur quelques espèces fossiles de l'ordre des thysanoptères. (Bull. soc. philom. Paris, (6), 10: 20-27.) 8°. Paris. 1873.
Describes a new genus and three new species of Physopoda from Aix.

——— Sur un hémiptère de la famille des pentatomides. (Bull. soc. philom. Paris, (6), 11: 14-16.) 8°. Paris. 1874 [1877].
Describes Cydnopsis heeri.

Schlechtendal, D. von. Physopoden aus dem braunkohlengebirge von Rott am Siebengebirge. (Zeitschr. ges. naturw., 60: 551-592, pl. 3-5.) 8°. Berlin, 1887.
Describes twelve species, referred to the genera Phloeothrips (1), Thrips (7), and Heliothrips (4).

Scudder, S. H. The tertiary Physopoda of Colorado. 8°. Washington. 1875. pp. 3. (Bull. U. S. geol. geogr. surv. terr., 1, ser. 2: 221-223.) 8°. Washington. 1875.
Describes two genera and three species.

——— Physiognomy of the American tertiary Hemiptera. (Proc. Bost. soc. nat. hist., 24: 562-579.) 8°. Boston. 1890.
Sets forth the proportional number of species in each of the families, and compares them with that of those now living in America and in the European tertiaries. The fauna of the tertiary west shown to bear definite relations to that now existing in the same region, but to have distinct tropical affinities. The total number of species is 266.

——— See also Buckton, G. B.

Signoret, Victor. See Millière, P.

VIIf.—Cenozoic Coleoptera.

. See also under Section I and Section VI.

Assmann, A. Palaeontologie. Beiträge zur insekten-fauna der vorwelt.—Einleitung. I. Beitrag. Die fossilen insekten des tertiären (miocenen) thonlagers von Schossnitz bei Kanth in Schlesien. II. Beitrag. Fossile insekten aus der tertiären (oligocenen) braunkohle von Naumburg am Bober. Mit einer tafel

Assmann, A.—Continued.
abbildungen. 8°. Breslau. 1869. pp.
1-62, taf. 1. (Zeitschr. f. entom. des
vereins f. schles. insektenk., (2), 1.)

The second paper describes two Coleoptera.
See same title in Section I and Section VI.

Baily, William Hellier. Notice of plant
remains from beds interstratified with the
basalt in the county of Antrim. (Quart.
journ. geol. soc. Lond., 25: 162, 357-362,
pl. 14-15.) 8°. London, 1869.

Two elytra of beetles "of distinct species, resembling those of some of the smaller Carabidæ,"
are mentioned, pp. 359-360, as occurring in the
leaf-bed, and are figured, pl. 14, figs. 14-15; in the
explanation of the plate, pp. 361-362, they are
compared to Rhynchophora. See also Judd, J. W.

Barrois, Jules. See Debray, H.

Barthélemy de la Lapommeraye, A.
Carabe d'Agassiz, Carabus agassizi. 8°.
pp. 4. Marseille. [1850.]

Extract and notice signed G. M. (Guérin Méne-
ville] under same title. (Rev. mag. zool., (2), 3:
203-204.) 8°. Paris. 1851.

Box. [Title of paper unknown.]
(Ann. rep. roy. inst. Cornwall, 26:—) 8°.
Truro. 1844.

Not seen. Notices elytra of beetles in a layer of
sand with vegetable matter beneath a marsh on
Millenbreath beach, Cornwall. Cf. Ussher (Geol.
mag., (2), 6: 251.), 8°. London. 1879.

Brodie, P. B. On the occurrence of
the remains of insects in the tertiary
clays of Dorsetshire. (Quart. journ. geol.
soc. Lond., 9, proc., 53-54.) 8°. London.
1853.

Refers to a few Curculionidæ and Buprestidæ
found at Corfe, afterwards figured by Westwood.

Brown, John. Insects and seeds in
peat at Stanway. (Geologist, 1858, 254.)
8°. London. 1858.

Notes the occurrence of elytra.

Curtis, J. See Lyell, C.

Debray, Henri. Tourbières du littoral
flamand et du département de la Somme.
(Bull. soc. géol. France, (3), 2: 46-49.)
8°. Paris. 1874.

Records the discovery, p. 48, of brilliant elytra
of Donacia in peat along the Flemish coast.

—— Communications diverses au
sujet des tourbières: castors; ossements
de baleines; élytres de donacies; sque-
lette humain des tourbières d'Aveluy;
crânes; bois. (Ann. soc. géol. nord, 5:
125-135.) 8°. Lille. 1876.

Debray, Henri—Continued.
Under the heading Insectes, refers to the discovery in peat at Ardres of Donacia sericea, determined by J. Barrois, pp. 127-128. Separately
issued, without change of pagination but with a
title-page. 8°. Lille. 1878.

Debray, L. Étude géologique et archéo-
logique de quelques tourbières du littoral flamand et du département de la
Somme. (Mém. soc. sc. agric. arts Lille,
(3), 11: 433-487, pl. 13.) 8°. Lille. 1873.

Contains a brief paragraph, p. 451, on the few
beetles found.

Desmarest, Eugène. Un morceau de
bois fossile . . . qui . . . a présenté des
traces qui ont dû être faites par des larves
d'insectes. (Ann. soc. ent. France, (2), 3,
bull., 26-27.) 8°. Paris. 1845.

Wood bored by larvæ of a longicorn beetle.

Desmoulins, Antoine. Découverte
d'élytres fossiles de coléoptères. (Ferr.,
Bull. sc. nat., 9: 253.) 8°. Paris. 1826.

Note upon a locality in the roche calcaire of
Mont St. Catherine, near Rouen, where elytra
with metallic colors had been found.

Douglas, John William, et al. A species
of stylopidæ fossil in amber. (Entom.
monthl. mag., 14: 18-19.) 8°. London.
1877.

A notice by the editors of the magazine of
Menge's Triæna.

Evans, C. E. Insect remains in the
Paludina beds at Peckham (with note
concerning them by F. Smith, as recorded in a letter from H. Woodward).
(Geologist, 4: 39-40, fig.) 8°. London.
1861.

Figures an elytron and mentions others.

Fisher, Osmond. On the brick-pit at
Lexden, near Colchester. (Quart. journ.
geol. soc. Lond., 19: 393-400.) 8°. London. 1863.

Under the head of Organic remains, pp. 398-400,
a letter is printed from T. V. Wollaston concerning Coleoptera found in the pit, and deductions
are drawn concerning the climate of the time in
which they lived.

Flach, K. Die käfer der unterpleisto-
cänen ablagerungen bei Hösbach unweit
Aschaffenburg. 8°. Würzburg. 1884.
(2), 13 pp., 2 pl. (Vorhandl. phys.-med.
gesellsch. Würzb., n. f., 18: 285-297,
pl. 8-9.) 8°. Würzburg. 1884.

Twenty-five species mostly Carabidæ and Chry-
somelidæ are described and excellently figured.

Fliche, P. Sur les lignites quaternaires de Jarville près de Nancy.

Fliche, P.—Continued.

(Comptes rendus, 80: 1233–1236.) 4°. Paris. 1875.

Records. p. 1234, seven kinds of beetles, northern species affecting moist localities, p. 1236. The insects were determined by Matthieu of the École forestière.

—— Faune et flore des tourbières de la Champagne. (Comptes rendus, 82: 979–982.) 4°. Paris. 1876.

Notices the occurrence, p. 979, of four species of beetles from Vannes.

Fritsch, A. Uiber [sic] einen fossilen maikäfer (Anomalites fugitivus Fr.) aus dem tertiären süsswasserquarz von Nogent le Rotrou in Frankreich. 8°. Prag. 1884. 3 pp., fig. (Sitzungsh. k. böhm. gesellsch. wiss. Prag, 1884, 163–165.) 8°. Prag. 1884.

Describes and figures a beetle found inclosed in a tertiary quartz pebble.

Früh, J. Kritische beiträge zur kenntniss des torfes. (Jahrb. k. k. geol. reichsanst., 35: 677–726, taf. 12.) 8°. Wien. 1885.

Mentions a Donacia on p. 679.

Gaudin, C. T. See **Heer, O.**

Giard, A. Les coléoptères fossiles d'Auvergne par M. Oustalet; remarques critiques. Bull. scient. dép. Nord, (2), 1: 56–62, 109–118.) 8°. Lille. 1878.

A sharp criticism of Oustalet's memoir on the fossil insects of Auvergne.

See same title in Section VIIc.

Gravenhorst, J. L. K. Monographia coleopterorum micropterorum. 16°. Gottingæ. 1805. pp. 16, 236, (12), tab. 1.

Contains, pp. 235–236, description of a single species of Oxyporus from amber, which in p. (3) of Index is given the specific name blumenbachii.

Grote, Augustus Radcliffe. Book notice. [Review of Scudder's contribution to Zittel's Handb. d. palæont.] (Can. ent., 18: 100.) 8°. London. 1886.

Proposes Mengea for Triæna preoccupied.

Guérin-Méneville, F. E. See **Barthélemy-Lapommeraye, A.**

Hammerschmidt, Karl Eduard. Neue käfer in bernstein. (Haidinger, Bericht fr. naturw. Wien, 1: 39.) 8°. Wien. 1847.

Mere exhibition of a specimen.

Heer, O. Ueber die vorweltlichen käfer von Oeningen. (Mittheil. naturf. gesellsch. Zürich. 1: 17–18.) 8°. Zürich. 1847.

A brief general statement of the peculiarities of the beetle-fauna of Oeningen.

Heer, O.—Continued.

—— Ueber die fossilen pflanzen von St. Jorge in Madeira. (Neue denkschr. allg. schweiz. gesellsch. gesammt. naturw., 15, art. 2.) 4°. Zürich. 1857. pp. 40, pl. 3.

Laparocerus wollastoni described in a note on p. 14, and figured pl. 2, fig. 34.

—— Les charbons feuilletés de Durnten et d'Utznach; discours de M. le professeur O. Heer traduit par M. Charles-Th. Gaudin. (Arch. sc. phys. nat., (n. p.), 2: 305–339.) 8°. Genève. 1858.

In a note, p. 322, mentions the occurrence of species of Donacia and Hylobius in the Durnten clays. This appears to be the only publication of the address.

—— Ueber die fossilen calosomen. 4°. [Zürich, 1860.] pp. 10, pl.

Published in the Programm of the Polytechnicum of Zürich. Seven species are described and figured from Locle and Oeningen, preceded by general remarks on fossil and recent Carabidæ.

—— Beiträge zur insektenfauna Oeningens. Coleoptera — geodephagen, hydrocanthariden, gyriniden, brachelytren, clavicornen, lamellicornen und buprestiden. (Natuurk. verhand. holl. maatsch. wetensch. Haarl., (2), 16: 1–90, taf. 1–7.) 4°. Haarlem. 1862.

Describes and figures 110 species, nearly all of them new. In an introduction of five pages some general results of the study of Oeningen Coleoptera are tabulated, the most interesting of which appear to be that the fauna is more European in character than the flora and less rich in American forms, and that many species are related to those which now enjoy a wide distribution.

—— Flora fossilis alaskana. Fossile flora von Alaska. 4°. Stockholm. 1869. pp. 41, pl. 10. (Kongl. svenska vetensk.-akad. handl., 8, iv.) 4°. Stockholm. 1869.

Describes Chrysomelites alaskanus, p. 39, pl. 10. Forms vol. 2, no. 3, of Heer's Flora fossilis arctica.

—— Nachträge zur miocenen flora Grönlands, enthaltend die von der schwedischen expedition im sommer 1870 gesammelten miocenen pflanzen. 4°. Stockholm. 1874. pp. 29, pl. 5. (Kongl. svenska vetensk.-akad. handl., 13, ii.)

Insekten, p. 25, pl. 5, describes two species of Clatelles.

Forms vol. 3, no. III, of Heer's Flora fossilis arctica.

—— Notes on fossil plants discovered in Grinnell Land by Captain H. W. Feilden, naturalist of the English north

Heer, O.—Continued.

polar expedition. (Quart. journ. geol. soc. Lond., 34: 66-72.) 8°. London. 1878.

Mentions, p. 69, the occurrence of a single elytron of a beetle with the plants.

—— Die miocene flora des Grinnell-Landes gegründet auf die von Cap. H. W. Feilden und Dr. E. Moss in der nähe des Kap Murchison gesammelten fossilen pflanzen. 4°. Zürich. 1878. pp. 48, front., pl. 9.

Describes and figures a single beetle.

Forms vol. 5, no. 1, of Heer's Flora fossilis arctica.

—— Primitiae florae fossilis sachaliuensis. Miocene flora der insel Sachalin. t. p., pp. 61, pl. 15. (Mém. acad. imp. sc. St.-Pétersb., 25, vii.) 4°. St.-Pétersbourg. 1878.

Describes a single beetle.

Forms vol. 5, no. iii, of Heer's Flora fossilis arctica.

—— See also **La Harpe, P. de.**

Henwood, William Jory. Observations on the detrital tin-ore of Cornwall. (Journ. roy. inst. Cornwall, 4: 191-254.) 8°. Truro. 1873.

Refers to the discovery of elytra of beetles in alluvium at Perranwell, Cornwall.

TRANSLATION: Remarques sur le minéral d'étain détritique du Cornwall. Traduction, par extraits, par Zeiller. (Ann. des mines, (7), 6: 114-130.) 8°. Paris. 1874.

The translation omits all references to animal remains.

Heyden, C. von. Fossile insekten aus der braunkohle von Sieblos. Nachtrag. (Palaeontogr., 8: 15-17, pl. 3, figs. 7-9.) 4°. Cassel. 1859.

Description of a beetle and two Hemiptera.

See same title in Section VIIe.

Heyden, C. und L. von. Käfer und polypen aus der braunkohle des Siebengebirges. (Palaeontogr., 15: 131-159 [Käfer, 131-157], pl. 22-24.) 4°. Cassel. 1866.

With another paper, reprinted under the title: Käfer und polypen aus der braunkohle des Siebengebirges.—Dipteren-larve aus dem tertiär-thon von Nieder-Flörsheim in Rhein-Hessen, mit 3 tafel abbildungen. Besonderer abdruck aus den Palaeontographicis, 15. 4°. Cassel. 1866. pp. 1-29, pl. 1-3.

Describes sixty beetles.

See same title in Section VIIg.

Hinde, George Jennings. The glacial and interglacial strata of Scarboro'

Hinde, G. J.—Continued.

Heights, and other localities near Toronto, Ontario. (Can. journ. sc. lit. hist., n. s., 15: 388-413, pl.) 8°. Toronto. 1877.

Mentions, p. 399, elytra of Carabidae in interglacial clays.

Hislop, S. See **Murray, A.**

Hollingworth, George H. Description of a peat-bed interstratified with the boulder drift at Oldham. (Quart. journ. geol. soc. Lond., 37: 713-714, fig.) 8°. London. 1881.

Reports beetles in the main bed of peat, p. 713.

Hope, F. W. Description de quelques insectes non décrits trouvés dans la résine animée. (Mag. de zool., (2), 4, Ins., pl. 87-89.) 8°. Paris. 1842.

Three coleoptera are described and figured in detail.

Horn, George Henry. Notes on some coleopterous remains from the bone cave at Port Kennedy, Pennsylvania. (Trans. Amer. ent. soc., 5: 241-245.) 8°. Philadelphia. 1876.

Collected, without change of pagination, with other papers under the title: Miscellaneous papers on American coleoptera. Eleven species are described.

Judd, John Wesley. The secondary rocks of Scotland. Second paper. On the ancient volcanoes of the Highlands and the relation of their products to the mesozoic strata. (Quart. journ. geol. soc. Lond., 30: 220-301, pl. 22-23.) 8°. London. 1874.

Mentions, p. 271, the discovery of "elytra of two species of beetles" in lacustrine deposits at Ballypalidy, Co. Antrim, Ireland, which he refers to the miocene. These beetles were figured by Bally (q. v.).

Kendall, J. D. Interglacial deposits of West Cumberland and North Lancashire. (Quart. journ. geol. soc. Lond., 37: 29-39, pl. 3.) 8°. London. 1881.

Records the occurrence of elytra of beetles in deposits at Drigg, p. 34, and St. Bees, p. 35, without mention of names.

Kerr, Washington Carothers. Report of the geological survey of North Carolina. Vol. i. Physical geography, résumé, economical geology. 1875. 8°. Raleigh. 1875. pp. 18, 325, 120, map, pl. (1), 8.

Reports on p. 157 the occurrence of "numerous shining wing-covers of beetles" in peat near Morganton, N. C.

Kolbe, H. J. Zur kenntniss von insektenbohrgängen in fossilen hölzern. (Zeitschr. deutsch. geol. gesellsch., 40: 131-137, pl. 11.) 8°. Berlin. 1888.

Describes borings of four different tertiary beetles, of which three are named, one from Lebanon, the others from Nieder Lausitz. A list of known similar borings is added.

La Harpe, Philippe de. Sur un gisement de tourbe glaciaire trouvé à Lausanne. (Bull. soc. vaud. sc. nat., 14: 456-458.) 8°. Lausanne. 1876.

Contains a letter from Heer, who examined the organic remains in the peat and found the elytra of a Donacia.

Landgrebe, Georg. Ueber einen im polir-schiefer des Habichts-waldes aufgefundenen käfer. (Neues jahrb. f. mineral., 1843: 137-142.) 8°. Stuttgart. 1843.

Describes an Aphodius.

[Linné, C. von.] Museum tessinianum, opera illustrissimi comitis, Dom. Car. Gust. Tessin. f°. Holmiae. 1753. pp. (8), 123, (9), pl. 14.

On p. 98 he enters Entomolithus coleoptri, unknown locality, which he likens to a carabid.

Lyell, C. On the boulder formation or drift, and associated fresh-water deposits composing the mud cliffs of eastern Norfolk. (Proc. geol. soc. Lond., 3: 171-179.) 8°. London. 1840.

Mention is made, p. 175, of the discovery of three elytra of Coleoptera, which Curtis determines to be identical with living British species of Donacia and Copris.

Matthieu. See **Fliche**, P.

Menge, A. See **Douglas**, J. W.

Mortillet, Gabriel. Description d'une nouvelle espèce de coléoptère fossile (Donacia genin) trouvé dans les lignites de Sonnaz. (Bull. soc. hist. nat. Savoie, 1850, p. 135.) (Les Alpes, 1850, no. 5.) (Arch. sc. phys. nat., 15: 78-79.) 8°. Genève. 1850.

Description of the species. Only the last reference seen.

Motschulsky, V. Die coleopterologischen verhältnisse und die käfer Russland's. 4°. Moskau, 1845. 131 pp., 1 pl. (Bull. soc. imp. nat. Mosc., 18: 1-131, tab. 1.) 8°. Moskau, 1845.

Contains brief memoranda on amber insects on pp. 98-100.

——— Genres et espèces d'insectes publiés dans différents ouvrages. Supplément au 6° vol. des Horae societatis

Motschulsky, V.—Continued. entomologicae rossicae. 8°. St. Pétersbourg. 1868. t. p., pp. 118.

Contains a list, p. 103, of Insectes contenus dans le succin described by the author.

Murray, Andrew. Notes on some fossil insects from Nágpur. (Quart. journ. geol. soc. Lond., 16: 182-185, 189, pl. 10, figs. 66-70.) 8°. London. 1860.

Occurs as a separate note in an article by Hislop, "On the tertiary deposits associated with the trap-rock in the East Indies, with descriptions of the fossil shells by the Rev. Stephen Hislop, and of the fossil insects by Andrew Murray, and a note on the fossil Cypridae by T. Rupert Jones." Thirteen Coleoptera (Buprestidae and Curculionidae) are mentioned and figured; only one is named.

Nöggerath, J. Notizen über fossile animalien. (Arch. ges. naturl., 2: 323-325.) 8°. Nürnberg. 1824.

Mentions the discovery of fossil beetles (p. 325), at Orsberg on the Rhine.

Oustalet, E. Recherches sur les insectes fossiles des terrains tertiaires de la France; deuxième partie. Insectes fossiles d'Aix en Provence. [Premier fascicule. Coléoptères d'Aix.] (Ann. sc. géol., 5, art. 2, pp. 1-347, pl. 1-6.) 8°. Paris. 1874. [pp. 1-136 and pl. 1-2, Febr. 15, 1874; pp. 137-347, pl. 3-6, May 15, 1874.] Also [with prem. partie] under the title: Recherches sur les insectes fossiles des terrains tertiaires de la France. Thèse présentée à la faculté des sciences. 8°. Paris. 1874. pp. 1-356, pl. 1-12.

The second part describes 81 species after the same plan as in the preceding memoir. 32 of them are Rhynchophora, 19 Staphylinidae, 11 Carabidae, and the rest scattered among various families; the memoir opens with a chapter of 74 pages on the geological relations of the gypsum beds of Aix. For first part, see Section VI.

——— See also **Giard**, A.

Peach, Charles William. An account of the fossil organic remains found on the south-east coast of Cornwall, and in other parts of that county. (Trans. roy. geol. soc. Cornwall, 6: 12-23.) 8°. Penzance. 1846.

Mentions the discovery of elytra of beetles in alluvial matter lying about the roots of trees resting on clay at Port Mellin, p. 23.

——— On the fossiliferous strata of part of the south-east coast of Cornwall. (Trans. roy. geol. soc. Cornwall, 7: 57-62, pl. 3-4.) 8°. Penzance. 1865.

Mentions, p. 62, the occurrence of elytra of beetles in similar situations as the preceding at Ready Money.

Ponzi, Giuseppe. Lavori degli insetti nelle lignite del Monte Vaticano. 4°. Roma. [1876.] pp. (3). (Atti reale accad. lincei, (2), 3: 375-377.) 4°. Roma. 1876.

Treats only of the same insect as the next.

—— I fossili del Monte Vaticano. 4°. Roma. 1876. pp. 37, tav. 3. (Atti r. accad. linc., (2), 3: 925-959, tav. 1-3.) 4°. Roma. 1876.

Under the name of Hyloblum tortonianum describes, p. 10 (932), and figures, pl. 1, fig. 9, borings in pine from pliocene deposits.

Robert, Eugène. Lettre sur les observations faites en Danemarck, en Norwége et en Suède, et dans laquelle il parle du succin, etc. (Bull. soc. géol. France, 9: 114-118.) 8°. Paris. 1838.

The letter is addressed to M. Cordier. The amber containing insects is mentioned on pp. 114-115. The insects mentioned are Aphodius fossor, Buprestis, Galeruca, "altises," and "lebouclier." all on p. 114. It is quoted in the Royal society's catalogue under the title: De la disposition de la tourbe à Elseneur et des insectes qu'on y trouve.

S. Fossil coleoptera. (Hardw. sc. gossip, 1867: 238.) 8°. London. 1867.

Describes finding four Coleoptera in a stratum of debituminised peat in a freestone quarry near Fifeness.

Schaufuss, L. W. Einige käfer aus dem baltischen bernsteine. (Berl. entom. zeitschr., 32: 266-270.) 8°. Berlin. 1888.

Describes three genera, each with one species.

Scheuchzer, J. J. Piscium querelae et vindiciae expositae. 4°. Tiguri. 1708. t. p., pp. 36, pl. 5.

Mentions, p. 15, and figures, pl. 2, a "Scarabæus in lapide fissili iulingensi" as a relic of the deluge.

Schilling, Peter Samuel. [Ueber den salz-bohrkafer, Ptinus salinus.] (Uebers. arb. veränd. schles. gesellsch. vaterl. kult., 1843: 174-175.) 4°. Breslau. 1844.

Describes beetles obtained from rock-salt in Wieliczka by dissolving the salt.

Scudder, S. H. Fossil Coleoptera from the Rocky Mountain tertiaries. (Bull. U. S. geol. geogr. surv. terr., 2: 77-87.) 8°. Washington. 1876.

Describes thirty species, mostly from Green River, Wyoming, and White River, Colorado.

—— Description of two species of Carabidæ found in the interglacial deposits of Scarboro' Heights, near Toronto, Canada. (Bull. U. S. geol. geogr. surv. terr., 3: 763-764.) 8°. Washington. 1877.

Referred to Loxandrus and Loricera.

Scudder, S. H.—Continued.

—— The operations of a prehistoric beetle. (Can. ent., 18: 194-196.) 8°. London. (Ont.) 1886.

Subcortical mines of a scolytid beetle found on a juniper root in interglacial clays.

—— Administrative report U. S. geological survey for the year 1886-1887. (Ann. rep. U. S. geol. surv., 8: 188-189.) 8°. Washington. 1889.

One hundred and twenty-six species of 63 genera of Carabidæ, Dytiscidæ, Hydrophilidæ, Silphidæ, Staphylinidæ, and Coccinellidæ have been found in the American tertiaries.

—— [Remains of Coleoptera in the interglacial clays of Scarboro', Ont.] (Proc. Bost. soc. nat. hist., 24: 467-468.) 8°. Boston. 1890.

Elytra of 32 species have been discovered, mostly Carabidæ (20 sp.) and Staphylinidæ (7 sp.). All are extinct, but resemble insects of the same region.

—— See also Grote, A. R.

Serres, P. M. T. de. Notice sur les cavernes à ossemens fossiles des carrières de calcaire grossier, situées aux environs de Lunel Vieil, dans le département de l'Hérault. (Mém. soc. linn. Paris, 5: 442-464.) 8°. Paris. 1827.

§ V. Insectes, p. 457, gives a very brief account of the remains of Coleoptera found in the cavern

Smith, Edward. On the stream works of Pentowan. (Trans. geol. soc. Lond., 4: 404-409.) 4°. London. 1817.

Records, p. 407, the "wings of coleopterous insects" at a depth of 40 feet from the surface of the ground below layers of peat and sea mud in a section at Pentowan, Cornwall, half a mile from the coast.

Smith, F. See Evans, C. E.

Stein, J. P. E. F. Miscellanea. A. Zwei bernstein käfer. (Berl. entom. zeitschr., 25: 221.) 8°. Berlin. 1881.

Describes two species of Bothrideres.

Unger, F. Die fossile flora von Szántó in Ungarn. (Denkschr. kais. akad. wiss. Wien, math.-nat. cl., 30, i: 1-20, pl. 1-5.) 4°. Wien. 1870.

Describes, pp. 3-4, and figures, pl. 1, figs. 13-14, the elytron and wing of a beetle, Melolonthites paraschlagiana.

Ussher, W. A. E. Pleistocene geology of Cornwall; part iv. Submerged forests and stream tin gravels. (Geol. mag., (2), 6: 251-263.) 8°. London. 1879.

Refers in several places, on the authority of others, to the discovery of coleopterous remains in pleistocene deposits.

Ussher, W. A. E.—Continued.

—— The post-tertiary geology of Cornwall. 8°. Hertford. 1879. pp. (4), 59, pl. (1). figs.

Refers, pp. 30, 32, to the occurrence of elytra in alluvium and clay at different localities, on the authority of others?

Weber, C. Otto. Der tertiärflora der niederrheinischen braunkohlenformation. (Palæontogr., 2: 115–236, pl. 18–25.) 4°. Cassel. 1851–1852.

Contains, pp. 225–250 (1852), Ueber ein fossiles torflager in der Vorder-Eifel bei dem dorfe Wohlscheid, in which, p. 229, pl. 25, figs. 17–18, a Pterostichus is noted.

Webster, Thomas. On the fresh-water formations in the Isle of Wight, with some observations on the strata over the chalk in the southeast part of England. (Trans. geol. soc. Lond., 2: 161–254.) 4°. London. 1814.

Mentions, pp. 194, 206, branches of trees found in the tertiary clays of Sheppey, bearing excrescences produced by insects; and, p. 230, the discovery of beetles in the tertiary deposits at Newport.

Westwood, J. O. See **Brodie, P. B.**

Wollaston, Thomas Vernon. Note on the remains of Coleoptera from the peat of Lexden brick-pit. (Quart. journ. geol. soc. Lond., 19: 400–401.) 8°. London. 1863.

Indicates briefly the generic affinities of a dozen species.

—— See also **Fisher, O.**

Woodward, H. See **Evans, C. E.**

.*. The statement is somewhere made that Debey, Deutsch. naturf. versamml., 1847: 269–323, or Rheinl. u. westf. verhandl., 1848: 113–425, has a reference to two kinds of Coleoptera at Aix; but neither has.

VIIg.—Cenozoic Diptera.

.*. See also under Section I and Section VI.

Breyn, J. P. Observatio de succinea globu, plantæ cujusdam folio imprægnata, rarissima. (Phil. trans., 34, 154–156, pl., fig. 2.) 4°. London. 1728.

Mentions a fly in amber, with minute figure of same. See same title in Section VIIb.

Brongniart, C. J. E. Observations sur un insecte fossile de la famille des diptères trouvé à Chadrat (Auvergne), (Protomyia oustaleti). (Ann. sc. géol., 7, art. 4, pp. 2.) 8°. Paris. 1876.

Simple description.

Brongniart, C. J. E.—Continued.

—— Note sur une nouvelle espèce de diptère fossile du genre Protomyia (P. oustaleti), trouvée à Chadrat (Auvergne). (Bull. soc. géol. France, (3), 4 : 459–460, pl. 13, figs. 5–6.) 8°. Paris. 1876.

Same as the preceding.

—— Note rectificative sur les espèces de bibionides fossiles du genre Plecia. (Bull. séances soc. ent. France, 1878, vi : 60–61.) 8°. Paris. 1878. (Ann. soc. ent. France, (5), 8, bull., 47–48.) 8°. Paris. 1878.

The fossil Diptera described as Protomyia and Bibiopsis belong to the modern genus Plecia.

—— Note rectificative sur quelques diptères tertiaires et en particulier sur un diptère des marnes tertiaires (miocène inférieur) de Chadrat (Auvergne) la Protomyia oustaleti qui devra s'appeler Plecia oustaleti. 8°. Lille. 1878. t. p. pp. 75–81. (Bull. scient. dép. nord, (2), ann. 1, pp. 75–81.) 8°. Lille. 1878.

Discusses in full the species of fossil Bibionidæ described as Protomyia and Bibiopsides, and concludes that all belong to Plecia; redescribes Plecia oustaleti; an enlargement of the preceding paper.

—— Note sur les tufs quaternaires de Bernouville près Gisors (Eure). 8°. Paris. 1880. pp. 3. (Bull. soc. géol. France, (3), 8 : 418–420.) 8°. Paris. 1880.

Records finding the larva of Stratiomys, p. 2 (419).

—— See also **Giard, A.**

Denton, William. On a mineral, resembling albertite, from Colorado. (Proc. Bost. soc. nat. hist., 10 : 305–306.) 8°. Boston. 1866.

The first account, p. 306 of fossil insects from the American tertiaries. He speaks only of Diptera in a petroleum shale.

Giard, A. Note sur un diptère nouveau pour la faune française (Penthetria holosericea Meig.) suivie de quelques remarques sur les bibionides fossiles. (Bull. scient. hist. litt. dép. nord, ann. 8, pp. 172–178.) 8°. Lille. 1876.

Discusses, pp. 177–178, the Penthetria vaillanti of Oustalet from Auvergne. Continued in the following.

—— Note sur les bibionides fossiles. (Bull. scient. dép. nord, (2), 1 : 12–16.) 8°. Lille. 1878.

Criticises the classification by Oustalet and Brongniart of various species placed by them in Protomyia. A continuation of the preceding.

—— See also **Oustalet, E.**

Heer, O. See Loew, H.

Heyden, C. von und L. von. Bibonidea aus der rheinischen braunkohle von Rott. (Palaeontogr., 11: 19-30, pl. 8, 9, figs. 1-12.) 4° [Cassel.] 1865.

Description of twenty-three species, mostly Protomyia, and remarks on three or four others.

—— Dipteren-larve aus dem tertiärthon von Nieder-Flörsheim in Rhein-Hessen. (Palaeontogr., 15: 157, pl. 23, fig. 22.) 4°. Cassel. 1866.

With another paper reprinted under the title: Käfer und polypen aus der braunkohle des Siebengebirges. — Dipteren-larve aus dem tertiärthon von Nieder-Flörsheim in Rhein-Hessen, mit 3 tafel abbildungen. Besonderer abdruck aus den Palaeontographicis. 15. 4°. Cassel. 1866. pp. 1-29, pl. 1-3.

Description of Muscidites deperditus.
See same title in Sect. VII.f.

Heyden, L. von. Fossile dipteren aus der braunkohle von Rott im Siebengebirge. 4°. Cassel. 1870. t. p., pp. 2, 30, pl. 1-2. (Palaeontogr., 17: 237-266, pl. 44-45.) 4°. Cassel. 1870.

Describes forty-one species of seventeen genera, besides seven larvae of two different genera. In an appendix, pp. 265-266, a few details are given of other insects, and the collections in which they are found.

Langius, Carolus Nicolaus. Historia lapidum figuratorum Helvetiae, ejusque viciniae, in quâ non solùm enarrantur omnia eorum genera, species et vires, aeneisque tabulis repraesentantur, sed insuper adducuntur eorum loca nativa, in quibus reperiri solent, ut cuilibet facile sit eas colligere, modo adducta loca adire libeat. 8°. Venetiis. 1708. 2 t. p., pp. (26), 165, tab. 52.

A single "Musca" from Oeningen is figured on pl. 7, fig. 5, and mentioned on p. 30.

Loew, Hermann. Ueber den bernstein und die bernstein fauna. 4°. Berlin, 1850. pp. 44. (Progr. königl. realsch. Meseritz, 1850, pp. 1-44.) 4°. Meseritz. 1850.

Separate Berlin edition not seen; of the other, pp. 28-44 are occupied by a general systematic review of the amber Diptera, of which many new genera and species are indicated with brief or no description. More than 10,000 specimens were examined by Loew, and about 575 species indicated.

—— Beschreibung einiger neuen Tipularia terricola. (Linn. entom., 5: 385-406, tab. 2.) 8°. Berlin. 1851.

Treats, pp. 400-401, pl. 2, figs. 16-23, of the genus Toxorrhina and figures three amber species.

Loew, H.—Continued.

—— Ueber die dipteren fauna des bernsteins. 4. Königsberg. 1861. pp. 13. (Amtl. ber. versamml. deutsch. naturf., 35: 88-98.) 4°. Königsberg. 1861.

An important discussion of the problems suggested by a study of the Diptera of the Prussian amber, of which at this time 850 species were known to the author, and of which over 650, belonging to 101 genera, had been satisfactorily determined. These insects belong to a single district fauna, and represent only a fragment of that, viz: those low flying Diptera which love moist places sheltered from the wind. The generic types which existed in the amber period have probably been preserved down to our time. Of all living types North American Diptera, especially those found from lat. 35° to 40° most nearly resemble the amber fauna - next to these, those of Europe.

TRANSLATION: On the Diptera or two-winged insects of the amber-fauna. 8°, New Haven. 1864. pp. 20. (Amer. journ. sc., (3), 37: 305-324.) 8°. New Haven. 1864.

Translation by R. von Osten Sacken, who adds a single brief note on living species common to Europe and America.

—— Monographs of the Diptera of North America; prepared for the Smithsonian institution. Part 1; edited, with additions, by R. Osten Sacken. 8°. Washington. 1862. pp. 24, 221, pl. 2.

References to amber Diptera, partly original, will be found on pp. 11, 17.

—— The same: Part 2. On the North American Dolichopodidae. 8°. Washington. 1864. pp. 11, 360, pl. 3-7.

A paragraph in his Supplement, pp. 321-322, points out that this family of American flies "shows the most remarkable analogy to the remains of the fossil fauna of the same family preserved in amber."

—— Berichtigung der generischen bestimmung einiger fossilen dipteren. (Zeitschr. gesamml. naturw., 32: 180-191, taf. 5.) 8°. Berlin. 1868.

A revision of the tertiary Bibionidae described by Heer.

Meyer, C. E. H. von. Fische und insekten der braunkohle bei Westerburg in Nassau. (Neues jahrb. mineral., 1851: 677.) 8°. Stuttgart. 1851.

Brief notice of the discovery of two flies.

Osten Sacken, Carl (Robert Romanoff) von. New genera and species of North American Tipulidae with short palpi, with an attempt at a new classification of the

Osten Sacken, C.—Continued.

tribe. (Proc. acad. nat. sc. Philad., 1859 ; 197–256, pl. 3–4.) 8°. Philadelphia. 1859.

Refers, pp. 200, 221, 251, to the relationship of Protoplasa, Elephantomyia and Rhamphidia to the species of the Baltic amber, and the identity of Toxorrhina and Limnobiorhynchus.

———— Appendix to the paper entitled New genera and species of North American Tipulidæ with short palpi, etc. (Proc. acad. nat. sc. Philad., 1860 ; 15–17.) 8°. Philadelphia. 1860.

Brief remarks on the amber genera Toxorrhina and Macrochile, p. 17.

———— Monographs of the Diptera of North America, Part 4. On the North American Tipulidæ. 8°. Washington. 1869. pp. 11, 345, pl. 4.

Compares the American fauna to that of the European amber fauna, pp. 37–38 ; devotes a couple of paragraphs, pp. 107–109, 112–114, to show that the amber species referred by Loew to Toxorrhina belong to Elephantomyia ; and another to the Eriocera of Baltic amber, pp. 251–252.

———— Ueber einige merkwürdigen fälle der geographischen verbreitung von Tipuliden. (Entom. nachr., 6 : 67–63.) 8°. Putbus. 1880.

Abstract of a paper published in the Tageblatt of the 52° Versammlung deutscher naturforscher, pp. 232–233. Contains a few words about the Tipulidæ of amber as compared with those living in N. America.

———— A relic of the tertiary period in Europe, Elephantomyia, a genus of Tipulidæ. (Mitth. münch. entom. ver., 5 : 152–154.) 8°. München. 1881.

Three species are found in amber, one in Europe and America and one in South Africa. Not seen. Noticed in Arch. sc. phys. nat., (3), 7 : 503. 8°. Genève. 1882.

———— See also **Loew, H.**

Oustalet, E. Note sur une empreinte de diptère fossile des marnes du gypse des environs de Paris. (Bull. soc. philom. Paris, (6), 9 : 161.) 8°. Paris. 1872.

Not seen ; title received from the author. Biblio chapuisi is described.

———— Réclamation sur une question de nomenclature. (Bull. séances soc. entom. France, 1878, vii : 72.) 8°. Paris. 1878. (Ann. soc. entom. France, (5), 8, bull., 60–61.) 8°. Paris. 1878. (Bull. scient. dép. du Nord, (2), 1 : 105–106.) 8°. Lille. 1878. (With notes by Giard.)

A claim that the name of the original describer of the species of Protomyia referred by Brongni-

Oustalet, E.—Continued.

art to Plecia should still remain attached to them. Giard refers to the opinion of Loew regarding Heer's Protomyia.

———— See also **Giard, A.**

Staub, Moriz. Tertiäre pflanzen von Felek bei Klausenburg. 8°. Budapest. 1883. t. p., 19 pp., 1, pl. (Mitth. jahrb. k. ung. geol. anst., 6 : 263–281, pl. 18.) 8°. Budapest. 1883.

Describes and figures Biblio kochii.

Unger, F. Fossile insecten. 4°. [Breslau.] 1841. pp. 16, pl. 2. (Acta acad. cæs. leop.-carol., 19, ii : 413–428, tab. 71–72.) 4°. Vratislaviæ et Bonnæ. 1842.

Describes and figures eight Diptera from the tertiary beds of Radoboj. A geological section is given on the first plate, and the larger part of the paper, pp. 415–424, is given to an account of the locality.

Williston, Samuel Wendell. Some interesting new Diptera. (Trans. Conn. acad. arts sc., 4 : 243–246, fig.) 8°. New Haven. 1880.

In describing a new nemestrid from Washington Territory, he discusses the fossil Palembolus of Florissant.

———— Synopsis of the North American Syrphidæ. 8°. Washington. 1886. pp. 30, 335, pl. 12. (Bull. U. S. nat. mus., 31.)

Contains, pp. 281–283, a section on Geological distribution, mostly a review of Florissant Syrphidæ.

VIIh.—Cenozoic Lepidoptera.

. See also under Section I and Section VI.

Boisduval, Jean Alphonse. Compte verbal du rapport . . . sur un dessin . . . qui représente une empreinte de lépidoptère fossile, trouvée dans les environs d'Aix. (Ann. soc. ent. France, 8, bull. ent., 11–12.) 8°. Paris. 1839.

Compares the butterfly to the modern genus Cyllo. Reproduced in Scudder's Fossil butterflies, p. 15.

———— Rapport sur une empreinte de lépidoptère trouvée dans les marnes des environs d'Aix en Provence, et communiquée par M. de Saporta. (Ann. soc. ent. France, 9 : 371–374, pl. 8.) 8°. Paris. 1840.

Description of Cyllo sepulta. Reproduced in Scudder's Fossil butterflies, pp. 15–17.

———— Quelques mots de réponse à M. Alex. Lefebvre sur les observations rela-

Boisduval, J. A.—Continued.
tives à la Cyllo sepulta, et à laquelle il donne pour épigraphe ces mots: Stupete gentes! (Ann. soc. ent. France, (2), 9, bull. ent., 96–98.) 8°. Paris. 1851.

Rejoinder maintaining his own position and ridiculing that of Lefebvre. Reproduced in Scudder's Fossil butterflies, p. 26.

Brodie, P. B. Fossil Lepidoptera. (Ann. rep. proc. Warw. nat. arch. field club, 1577, pp. 3–9.) 8°. Warwick. 1877.

Not seen. The author states that it is based on Scudder's work on the subject, and contains nothing original.

Butler, A. G. Catalogue of the diurnal lepidoptera of the family Satyridæ in the collection of the British museum. 8°. London. 1868. pp. 6, 211, pl. 5.

Under the heading Fossil species, pp. 189–190, are given brief notes on Neorinopis sepulta, to show that it "is exactly intermediate in character between . . . Neorina, Antirrhæa, and Anchyphlobia."

—— Catalogue of diurnal Lepidoptera described by Fabricius in the collection of the British museum. 8°. London. 1869. pp. 5, 303, pl. 3.

Refers, p. 109, to a possible relationship between Argynnis diana and the fossil Vanessa pluto.

—— Description of a new genus of fossil moths belonging to the geometrid family Euschemidæ. (Proc. zool. soc. Lond., 1889: 292–297, pl. 31.) 8°. London. 1889.

Describes Lithopsyche antiqua from the Isle of Wight tertiaries, figures it and related living forms, and discusses the development of color in Lepidoptera.

Daudet, Henri. Chenilles fossiles. (Petites nouv. entom., 2, no. 145, p. 25.) 4°. Paris. 1876.

First mention of the discovery of caterpillars at Aix.

—— Description d'une chenille fossile trouvée dans le calcaire d'Aix (Provence). (Rev. mag. zool., (3), 4: 415–424, pl. 17.) 8°. Paris. 1876.

Describes and figures Satyrites incertus, the first fossil caterpillar of a butterfly known, and discusses its probable affinities.

Doubleday, Edward, and Westwood, John Obadiah. The genera of diurnal Lepidoptera; comprising their generic characters, a notice of their habits and transformations, and a catalogue of the species of each genus; illustrated with 86 plates, by William C. Hewitson. 2 vols. fol. London. 1846–'52. Vol. 1,

Doubleday and Westwood.—Cont'd.
pp. 12, 250, (2), pl. A, 1–30:—vol. 2, t.p., pp. 251–534, pl. 31–80, and suppl. pl.

Several numbers on the plates are repeated, followed by "A." A single fossil species, Cyllo sepulta, is catalogued on p. 361.

Duponchel [Philogène Auguste Joseph]. L'existence d'une impression très-remarquable de lépidoptère fossile, qui a été trouvée dans une plâtrière des environs d'Aix (en Provence). (Ann. soc. ent. France, 7, bull. ent., 51–52.) 8°. Paris. 1838.

First announcement of Neorinopis as a "Nymphale." Reproduced in Scudder's Fossil butterflies, p. 15.

Edwards, William Henry. The butterflies of North America. 4°. New York. 1868–'72. pp. (10), 2, (154), pl. (50). Contains also a synopsis of North American butterflies, pp. 5, 52, and a supplementary part, pp. (17), pl. (3), and corrected pp. 4–12, 19–20 of synopsis.

P. (64) in pt. 1 (1868) contains a figure of Mylothrites pluto, with suggestions concerning its affinities with the living Argynnis diana.

[Gervais, P.] Fossiles du Quercy. (Journ. de zool., 6: 67–69.) 8°. Paris. 1877.

Records a pupa of Triphæna from tertiary deposits.

Goss, H. Fossil Lepidoptera. (Entom., 21: 66–67.) 8°. London. 1888.

A paragraph giving a general account of the known fossil butterflies.

Kawall, J. H. Organische einschlüsse im bergkrystall. (Bull. soc. imp. nat. Moscou, 1876, no. 3, pp. 170–173.) 8°. Moscou. 1876.

Describes a caterpillar, Tineites crystalli, found in quartz from Siberia.

Kirby, W. F. A synonymic catalogue of diurnal Lepidoptera. 8°. London. 1871. pp. 7, 690. Supplement, March, 1871–June, 1877. 8°. London. 1877. pp. 7, 691–883.

Includes the few fossil species.

Lefebvre, Alexandre. Observations relatives à l'empreinte d'un lépidoptère fossile (Cyllo sepulta) du Docteur Boisduval. (Ann. soc. ent. France, (2), 9: 71–88, pl. 3, ii.)

An argument to show that Boisduval had wrongly interpreted both the neuration and the markings of the wings. Reproduced in Scudder's Fossil butterflies, pp. 17–25, pl. 1, figs. 14–16.

Minot, Charles Sedgwick. Zur kennt-niss der insektenhaut. (Arch. mikr. anat., 28: 37–48, pl. 7.) 8°. Bonn, 1886.

Describes on pp. 46–47 the structure of the in tegument of a fossil caterpillar from the oligocene of Florissant.

Scudder, S. H. Description d'un nouveau papillon fossile (Satyrites rey-nesii) trouvé à Aix en Provence. 8°. Paris. 1872. pp. 7, pl. 1. (Rev. mag. zool., 1871–72: 66–72, pl. 7.)

EXTRACT: Description of a new fossil butterfly (Satyrites reynesii) found at Aix in Provence. 8°. London. 1872. pp. 2, pl. 1. (Geol. mag., 9: 532–533, pl. 13, figs. 2, 3.)

Also entitled on cover: On a new fossil butter-fly. The English translation is by the editor of the Geological magazine. The species is from the tertiary.

——— Fossil butterflies. (Mem. Amer. assoc. adv. sc., 1.) 4°. Salem. 1875, pp. 12, 99, pl. 3.

Describes in detail the generic and specific characters of the five known species, besides four new ones, all from European tertiaries. After an historical introduction there are sections on their geological relations, the probable food plants of their caterpillars, the present distribution of their nearest allies, and the fossils believed to be erro-neously referred to butterflies. More or less ex-tended abstracts will be found in Arch. sc. phys. nat., n. s., 55: 102–103; 57: 91–92. 8°. Genève. 1876.—Neues Jahrb. miner., 1877: 445–417. 8°. Stuttgart. 1877;—Amer. journ. sc., (3), 11: 74–75. 8°. N. Haven. 1876;—Amer. nat., 10: 53, 106–107. 8°. Salem. 1876. See also Boisduval, J. A.; Duponchel, P. A. J.; and Lefebvre, A.

——— Fossil butterflies. (Scudd., Butt. New Engl., 1: 756–760, pl. 16, fig. 6.) 8°. Cambridge. 1889.

Forms Excursus xxiv. A general account of the relations of the few species known in Europe and America.

——— ' The fossil butterflies of Floris-sant. (Ann. rep. U. S. geol. surv., 8: 433–474, pl. 52, 53.) 8°. Washington. 1889 [1890].

Descriptions and illustrations of seven species, with an appendix on a living African libythedd allied to one of them: each is referred to a dis-tinct and extinct genus and all but one are Nym-phalidæ.

——— See also Brodie, P. B.; Lefeb-vre, A.; and Strecker, H.

Serres, P. M. T. de. Deuxième note additionnelle au Mémoire géologique . . .

Serres, P. M. T. de.—Continued. sur la Provence. (Actes soc. linn. Bord., 13: 170–172.) 8°. Bordeaux. 1843.

Page 172 contains a Note relative au lépidop-tère figuré au no. 4; but the plate of the butterfly. Neorinopis sepulta, appears to have been pub-lished in a limited edition only, as the two or three copies I have examined do not contain it. Two years later Serres mentions its publication.

——— Sur les fossiles du bassin d'Aix (Bouches du Rhône). (Ann. sc. nat., (3), zool., 4: 249–256). 8°. Paris. 1845.

Pp. 251–254 are mostly given to combating the arguments drawn from the presence of Cyllo sepulta in favor of the equatorial nature of the ancient climate of Aix.

Strecker, Herman. Butterflies and moths in their connection with agricul-ture and horticulture: a paper prepared for the Pennsylvania fruit growers' society, January, 1879. 8°. Harrisburg. 1879. 22 pp.

Contains, p. 19, a paragraph on fossil butter-flies drawn from Scudder's paper, with one or two comments.

Swinton, A. H. A study of the vari-ation of the small tortoise-shell butterfly (Vanessa urticæ). (Hardw. science gos-sip, 1881: 147–149, 176–179, figs. 88, 104, 105.) 1 8°. London. 1881.

A study of the evolution and specialization of butterflies and moths, showing how the markings of the wings of fossil Lepidoptera harmonize with the systematic design found in recent species; in figs. 104, 105, on p. 177, he attempts restorations of Neorinopis sepulta and a tertiary Bombyx, both from Aix.

Westwood, J. O. See **Doubleday** E., and **Westwood,** J. O.

VIII.—Cenozoic Hymenoptera.

. See also under Section I and Sec-tion VI.

Andrae, K. J. Beiträge zur kenntnisse der fossilen flora Siebenbürgens und des Banates. Mit zwölf tafeln. pp. 1–48, pl. 1–12. (Abhandl. k. k. geol. reichsanst. Wien, bd. 2, abth. 3, no. 4.) 4°. Wien. 1855.

Figures a Formica, pl. 4, fig. 6, 6a, 8b, with mention on p. 26. The Formica is said to come from Thalheim, but the locality of the plant on the same slab with it is given as Sutzka!

See same title in Section VIIc.

Brischke, D. Die hymenopteren des bernsteins. 8°. Danzig. 1886. pp. 2.

Brischke, D.—Continued.
(Schrift. naturf. gesellsch. Danz., n. f., 6,
iii: 278–279.) 8°. Danzig. 1886.

Brief summary of the genera known, based in
part on Menge's collection, with the number of
specimens in each.

Duisburg, H. von. Zur bernstein-
fauna. (Schriften k. phys.-ökon. gesellsch. Königsb., 9: 23–28, fig.) 4°.
Königsberg. 1868.

Discusses the systematic position of the small-
est amber insect known, a species of the hymenopterous genus Myrmar, the expanse of whose
wings is scarcely more than half a millimeter.

Emery, C., et Forel, Auguste. Catalogue des formicides d'Europe. (Mitth.
schweiz. entom. gesellsch., 5: 441–481.)
8°. Schaffhausen. 1879.

Contains, p. 481, Liste des ouvrages traitant des
formicides fossiles (6 titles).

Faujas-de-Saint-Fond, Barthélemy.
Nouvelle notice sur les plantes fossiles
renfermées dans un schiste marneux des
environs de Chaumerac et de Roche
Sauve, département de l'Ardèche. (Mém.
mus. hist. nat., 2: 444–459, pl. 15.) 4°.
Paris. 1815.

Gives the opinion of Latreille on a species of
"Polistes" figured on the plate.

Forel, Auguste. See **Emery, C., et
Forel, Auguste.**

Haesbert, Martin Johann. De con-
chylio et ape petrifactis. (Ephem. med.
phys. acad. caes. leop. nat. curios., dec. 3,
ann. 2, pp. 4–49.) 4°. Leipzig. 1695.

Reports a fossil bee in the collection of Schei-
dius, figured tab. 2, fig. 4.

Heer, O. Ueber fossile ameisen. (Mittheil. naturf. gesellsch. Zürich, 1: 167–
174.) 8°. Zürich. 1848.

The fossil ants of Oeningen and Radoboj are
winged and either males or females, neuters being
rarely preserved; three fourths are females. The
individuals are very abundant, and are preserved
in large assemblages, and many species in close
contiguity. Most of them are Formicidae, and
they form the best data for comparison of the Oeningen and Radoboj faunas.

TRANSLATION: On fossil ants. (Quart.
journ. geol. soc. Lond., 6, ii: 61–65.) 8°.
London. 1870.

Translated by T. R. J[ones].

——— Fossile hymenopteren aus Oeningen und Radoboj. 4. [n. p. n. d.]
pp. 42, pl. 3. (Neue denkschr. allgem.
schweiz. gesellsch. gesammt. naturw.,
22.) 4°. Zürich. 1867.

Heer, O.—Continued.

Catalogues and describes sixty-nine species. In
an appendix, p. 42, notice is taken of Mayr's criticism of his former treatment of the fossil ants.

——— See also **Mayr, G. L.**

Latreille, Pierre André. See **Faujas-
de-Saint-Fond, B.**

Malfatti, G. Due piccoli imenotteri
fossili dell' ambra siciliana. 4°. [Roma.
1881.] pp. 4, figs. (Atti accad. linc., (3),
trans., 5: 80–83, 2 figs.) 4°. Roma. 1881.

Describes and figures a Myrmar and a Tapi-
noma.

Martialis, Marcus Valerius. Epigram-
mata. Liber 4, section 32.

Et latet et lucet Phaethontide condita gutta,
Ut videatur apis nectare clausa suo.
Dignum tantorum pretium tulit illa laborum:
Credibile est ipsam sic voluisse mori.

Some writers have thought that Martial here
referred to amber-inclusa.

Mayr, Gustav Leopold. Vorläufige
studien über die Radoboj-formiciden in
der sammlung der k. k. geologischen
reichsanstalt. 8°. Wien. 1867. pp. 16,
pl. 1. (Jahrb. geol. reichsanst., 17: 47–
61, taf. 1.) 8°. Wien. 1867.

A revision of the specimens described by Heer
with reference to modern genera. See also
Heer, O.

ABSTRACT: On fossil insects. (Quart.
journ. geol. soc. Lond., 23, ii: 7.) 8°.
London. 1867.

——— Die ameisen des baltischen
bernsteins; mit 106 figuren auf fünf tafeln. (Beitr. naturk. Preussens, 1) pp. 4,
102, (10), tab. 5. 4°. Königsberg. 1868.

Extended descriptions of forty-nine species and
twenty-three genera, with some preliminary
general observations, including a review of previous literature, and a comparison of amber
species with modern types and those of Radoboj.

Saussure, Henri de. Note sur un
nouvel insecte hyménoptère fossile. 8°.
[Paris. 1852.] pp. 2, pl. (Rev. mag.
zool., (2), 4: 579–582, pl. 23, figs. 5–6.)
8°. Paris. 1852.

Describes and figures Pimpla antiqua from tertiaries of Aix.

Scudder, S. H. Note on fossil ants
from South Park, Colorado. (Amer. nat.,
11: 191.) 8°. Salem. 1877.

About forty species had been found at Florissant.

VIII.—COPAL INSECTS.

.°. See also under Section I and Section VI.

Bloch, Mark Eliez. Baytrag zur naturgeschichte des kopals. 16⁰. Berlin. 1776. (Beschäft. berl. gesellsch. naturf. fr., 2 : 91–196, tab. 3–5.)

Contains, pp. 164–190. Verzeichniss einiger merkwürdigen insekten, welche in kopal eingeschlossen, with rude figures.

Dalman, Johann Wilhelm. Om insekter inneslutne i copal; jemte beskrifning på några deribland förekommande nya slägten och arter. 8⁰. Stockholm. 1826. (Kongl. vetensk.-akad. handl., 1825, 375–411, tab. 5.) 8⁰. Stockholm. 1826.

Describes several new genera and species of insects found in African gum copal. Separate copy not seen. See also Lucas, H.

ABSTRACT : Des insectes renfermés dans les résines de copal. (Ferr., Bull. sc. nat., 14 : 257–290.) 8⁰. Paris. 1828.

A very full abstract including descriptions of the species, signed D. (Desmarest?).

ABSTRACT : On insects enclosed in copal. (Quart. journ. sc. lit. arts, 1828, 227–228.) 8⁰. London. 1828.

Briefer abstract of same.

Desmarest, A. G. See Dalman, J. W.

Gistl, Johannes Nepomuk Franz Xaver. Kerfe in copal eingeschlossen. (Oken, Isis, 1831, 247–248.) 4⁰. Leipzig. 1831.

Describes four new species from Brazil. Noticed in Neues Jahrb. mineral., 1833, 712. 8⁰. Stuttgart. 1833.

Hagen, H. A. Synopsis of the Psocina without ocelli. (Entom. monthl. mag., 2 : 121–124.) 8⁰. London. 1865.

Atropos resinata from copal is described on p. 121.

—— On some aberrant genera of Psocina. (Ent. monthl. mag., 2 : 148–152, 170–172.) 8⁰. London. 1865–'66.

Describes three species from copal.
See same title in Section VIIc.

—— Psocinorum et embidinorum synopsis. 8⁰. Wien. 1866. pp. 22. (Verh. zool.-bot. gesellsch. Wien, 1866, 201–222.) 8⁰. Wien. 1866.

Includes three Psocina from copal.
See same title in Section VIIc.

Loew, H. Dipterologische beiträge [1. abtheilung]. (Öffentl. prüf. Friedr.-Wilh. gymn. Posen, 1845 : 1–52, pl.) 4⁰. Posen. 1845.

Contains descriptions and figures of three copal Diptera.

Lucas, Hippolyte. [Note sur les espèces de coléoptères décrites par Dalman dans son mémoire: Om insekter inneslutne i copal.] (Bull. séances soc. entom. France, 1878, vii : 75.) 8⁰. Paris. 1878.

Remarks on the omission of those insects from Gemminger and Harold's Catalogus coleopterorum, and cites the species of all orders figured by Dalman.

Quedenfeldt, G. Copal-insecten aus Africa. (Berl. entom. zeitschr., 29 : 363–365.) 8⁰. Berlin. 1885.

Describes three Coleoptera.

Raffray, Achille. Note sur des insectes renfermés dans des morceaux de gomme copal. (Ann. soc. ent. France, (5), 5, bull. ent., 125–126.) 8⁰. Paris. 1875.

General notes based on a collection exhibited. A single paragraph is devoted to insects.

Swagerman, Everard Pieter. Waarneeming omtrent de insekten, welkon in de gomlak gevonden worden. (Verhand. zeeuwsch. genootsch. wetensch. Vliss., 7, ii : 227–258, pl.) 8⁰. Middelburg. 1780.

On copal insects.

Westwood, J. O. Characters of Embia, a genus of insects allied to the white ants (Termites): with descriptions of the species of which it is composed. (Trans. linn. soc. Lond., 17 : 369–374, pl. 11.) 4⁰. London. 1837.

In a postscript on p. 374 two species from gum copal are noticed.

—— On the economy of the genus Palmon of Dalman with descriptions of several species belonging thereto. (Trans. entom. soc. Lond., 4 : 256–281, pl. 10 pars.) 8⁰. London. 1847.

Redescribes some of Dalman's copal insects. The living species, whose economy is known, are parasitic on eggs of Mantidæ.

INDEX OF AUTHORS.

99

Series title.

United States. *Department of the interior.* (*U. S. geological survey*).

Department of the interior | — | Bulletin | of the | United States | geological survey | no. 69 | [Seal of the department] |

Washington | government printing office | 1890

Second title: United States geological survey | J. W. Powell, director | — | A | classed and annotated | bibliography of fossil insects | by | Samuel Hubbard Scudder | [Vignette] |

Washington | government printing office | 1890

8°. 101 pp.

Author title.

Scudder (Samuel Hubbard).

United States geological survey | J. W. Powell, director | — | A | classed and annotated | bibliography of fossil insects | by | Samuel Hubbard Scudder | [Vignette] |

Washington | government printing office | 1890

8°. 101 pp.

[United States. *Department of the interior.* (*U. S. geological survey*). Bulletin 69].

Title for subject entry.

United States geological survey | J. W. Powell, director | — | A | classed and annotated | bibliography of fossil insects | by | Samuel Hubbard Scudder | [Vignette] |

Washington | government printing office | 1890

8°. 101 pp.

[United States. *Department of the interior.* (*U. S. geological survey*). Bulletin 69].

ADVERTISEMENT.

[Bulletin No. 69.]

The publications of the United States Geological Survey are issued in accordance with the statute approved March 3, 1879, which declares that—

"The publications of the Geological Survey shall consist of the annual report of operations, geological and economic maps illustrating the resources and classification of the lands, and reports upon general and economic geology and paleontology. The annual report of operations of the Geological Survey shall accompany the annual report of the Secretary of the Interior. All special memoirs and reports of said Survey shall be issued in uniform quarto series if deemed necessary by the Director, but otherwise in ordinary octavos. Three thousand copies of each shall be published for scientific exchanges and for sale at the price of publication; and all literary and cartographic materials received in exchange shall be the property of the United States and form a part of the library of the organization: And the money resulting from the sale of such publications shall be covered into the Treasury of the United States."

On July 7, 1882, the following joint resolution, referring to all Government publications, was passed by Congress:

"That whenever any document or report shall be ordered printed by Congress, there shall be printed, in addition to the number in each case stated, the 'usual number' (1,900) of copies for binding and distribution among those entitled to receive them."

Except in those cases in which an extra number of any publication has been supplied to the Survey by special resolution of Congress or has been ordered by the Secretary of the Interior, this office has no copies for gratuitous distribution.

ANNUAL REPORTS.

I. First Annual Report of the United States Geological Survey, by Clarence King. 880. 8°. 79 pp. 1 map.—A preliminary report describing plan of organization and publications.

II. Second Annual Report of the United States Geological Survey, 1880-'81, by J. W. Powell. 1882. 8°. lv, 588 pp. 62 pl. 1 map.

III. Third Annual Report of the United States Geological Survey, 1881-'82, by J. W. Powell. 1883. 8°. xviii, 564 pp. 67 pl. and maps.

IV. Fourth Annual Report of the United States Geological Survey, 1882-'83, by J. W. Powell. 1884. 8°. xxxii, 473 pp. 85 pl. and maps.

V. Fifth Annual Report of the United States Geological Survey, 1883-'84, by J. W. Powell. 1885. 8°. xxxvi, 469 pp. 58 pl. and maps.

VI. Sixth Annual Report of the United States Geological Survey, 1884-'85, by J. W. Powell. 1885. 8°. xxix, 570 pp. 65 pl. and maps.

VII. Seventh Annual Report of the United States Geological Survey, 1885-'86, by J. W. Powell. 1888. 8°. xx, 650 pp. 71 pl. and maps.

VIII. Eighth Annual Report of the United States Geological Survey, 1886-'87, by J. W. Powell. 1889. 8°. 2 v. xix, 474, xli pp. 53 pl. and maps; 1 p. l. 475-1063 pp. 54-76 pl. and maps.

IX. Ninth Annual Report of the United States Geological Survey, 1887-'88, by J. W. Powell. 1889. 8°. xiii, 717 pp. 88 pl. and maps.

The Tenth and Eleventh Annual Reports are in press.

MONOGRAPHS.

I. Lake Bonneville, by Grove Karl Gilbert. 1890. 4°. xx, 438 pp. 51 pl. 1 map. Price $1.50.

II. Tertiary History of the Grand Cañon District, with atlas, by Clarence E. Dutton, Capt. U. S. A. 1882. 4°. xiv, 264 pp. 42 pl. and atlas of 24 sheets folio. Price $10.12.

III. Geology of the Comstock Lode and the Washoe District, with atlas, by George F. Becker. 1882. 4°. xv, 422 pp. 7 pl. and atlas of 21 sheets folio. Price $11.00.

IV. Comstock Mining and Miners, by Eliot Lord. 1883. 4°. xiv, 451 pp. 3 pl. Price $1.50.

I

13. Boundaries of the United States and of the several States and Territories, with a Historical Sketch of the Territorial Changes, by Henry Gannett. 1885. 8°. 135 pp. Price 10 cents.

14. The Electrical and Magnetic Properties of the Iron-Carburets, by Carl Barus and Vincent Strouhal. 1885. 8°. 238 pp. Price 15 cents.

15. On the Mesozoic and Cenozoic Paleontology of California, by Charles A. White. 1885. 8°. 33 pp. Price 5 cents.

16. On the Higher Devonian Faunas of Ontario County, New York, by John M. Clarke. 1885. 8°, 86 pp. 3 pl. Price 5 cents.

17. On the Development of Crystallization in the Igneous Rocks of Washoe, Nevada, with Notes on the Geology of the District, by Arnold Hague and Joseph P. Iddings. 1885. 8°. 44 pp. Price 5 cents.

18. On Marine Eocene, Fresh-water Miocene, and other Fossil Mollusca of Western North America, by Charles A. White. 1885. 8°. 26 pp. 3 pl. Price 5 cents.

19. Notes on the Stratigraphy of California, by George F. Becker. 1885. 8°. 28 pp. Price 5 cents.

20. Contributions to the Mineralogy of the Rocky Mountains, by Whitman Cross and W. F. Hillebrand. 1885. 8°. 114 pp. 1 pl. Price 10 cents.

21. The Lignites of the Great Sioux Reservation. A Report on the Region between the Grand and Moreau Rivers, Dakota, by Bailey Willis. 1885. 8°. 16 pp. 5 pl. Price 5 cents.

22. On New Cretaceous Fossils from California, by Charles A. White. 1885. 8°. 25 pp. 5 pl. Price 5 cents.

23. Observations on the Junction between the Eastern Sandstone and the Keweenaw Series on Keweenaw Point, Lake Superior, by R. D. Irving and T. C. Chamberlin. 1885. 8°. 124 pp. 17 pl. Price 15 cents.

24. List of Marine Mollusca, comprising the Quaternary Fossils and recent forms from American Localities between Cape Hatteras and Cape Roque, including the Bermudas, by William Henley Dall 1885. 8°. 336 pp. Price 25 cents.

25. The Present Technical Condition of the Steel Industry of the United States, by Phineas Barnes. 1885. 8°. 85 pp. Price 10 cents.

26. Copper Smelting, by Henry M. Howe. 1885. 8°. 107 pp. Price 10 cents.

27. Report of work done in the Division of Chemistry and Physics, mainly during the fiscal year 1884-'85. 1886. 8°. 80 pp. Price 10 cents.

28. The Gabbros and Associated Hornblende Rocks occurring in the Neighborhood of Baltimore, Md. by George Huntington Williams. 1886. 8°. 78 pp. 4 pl. Price 10 cents.

29. On the Fresh-water Invertebrates of the North American Jurassic, by Charles A. White. 1886 8°. 41 pp. 4 pl. Price 5 cents.

30. Second Contribution to the Studies on the Cambrian Faunas of North America, by Charles Doolittle Walcott. 1886. 8°. 369 pp. 33 pl. Price 25 cents.

31. Systematic Review of our Present Knowledge of Fossil Insects, including Myriapods and Arachnids, by Samuel Hubbard Scudder. 1886. 8°. 128 pp. Price 15 cents.

32. Lists and Analyses of the Mineral Springs of the United States; a Preliminary Study, by Albert C. Peale. 1886. 8°. 235 pp. Price 20 cents.

33. Notes on the Geology of Northern California, by J. S. Diller. 1886. 8°. 23 pp. Price 5 cents.

34. On the relation of the Laramie Molluscan Fauna to that of the succeeding Fresh-water Eocene and other groups, by Charles A. White. 1886. 8°. 54 pp. 5 pl. Price 10 cents.

35. Physical Properties of the Iron-Carburets, by Carl Barus and Vincent Strouhal. 1886. 8°. 62 pp. Price 10 cents.

36. Subsidence of Fine Solid Particles in Liquids, by Carl Barus. 1886. 8°. 58 pp. Price 10 cents.

37. Types of the Laramie Flora, by Lester F. Ward. 1887. 8°. 354 pp. 57 pl. Price 25 cents.

38. Peridotite of Elliott County, Kentucky, by J. S. Diller. 1887. 8°. 31 pp. 1 pl. Price 5 cents.

39. The Upper Beaches and Deltas of the Glacial Lake Agassiz, by Warren Upham. 1887. 8°. 84 pp. 1 pl. Price 10 cents.

40. Changes in River Courses in Washington Territory due to Glaciation, by Bailey Willis. 1887. 8°. 10 pp. 4 pl. Price 5 cents.

41. On the Fossil Faunas of the Upper Devonian—the Genesee Section, New York, by Henry S. Williams. 1887. 8°. 121 pp. 4 pl. Price 15 cents.

42. Report of work done in the Division of Chemistry and Physics, mainly during the fiscal year 1885-'86. F. W. Clarke, chief chemist. 1887. 8°. 152 pp. 1 pl. Price 15 cents.

43. Tertiary and Cretaceous Strata of the Tuscaloosa, Tombigbee, and Alabama Rivers, by Eugene A. Smith and Lawrence C. Johnson. 1887. 8°. 189 pp. 21 pl. Price 15 cents.

44. Bibliography of North American Geology for 1886, by Nelson H. Darton. 1887. 8°. 35 pp. Price 5 cents.

45. The Present Condition of Knowledge of the Geology of Texas, by Robert T. Hill. 1887. 8°. 94 pp. Price 10 cents.

46. Nature and Origin of Deposits of Phosphate of Lime, by R. A. F. Penrose, jr., with an Introduction by N. S. Shaler. 1888. 8°. 143 pp. Price 15 cents.

47. Analyses of Waters of the Yellowstone National Park, with an Account of the Methods of Analysis employed, by Frank Austin Gooch and James Edward Whitfield. 1888. 8°. 84 pp. Price 10 cents.

STATISTICAL PAPERS.

Mineral Resources of the United States, 1882, by Albert Williams, jr. 1883. 8°. xvii, 813 pp. Price 50 cents.

Mineral Resources of the United States, 1883 and 1884, by Albert Williams, jr. 1885. 8°. xiv, 1016 pp. Price 60 cents.

Mineral Resources of the United States, 1885. Division of Mining Statistics and Technology. 1886. 8°. vii, 576 pp. Price 40 cents.

Mineral Resources of the United States, 1886, by David T. Day. 1887. 8°. viii, 813 pp. Price 50 cents.

Mineral Resources of the United States, 1887, by David T. Day. 1888. 8°. vii, 832 pp. Price 50 cents.

Mineral Resources of the United States, 1888, by David T. Day. 1890. 8°. vii, 652 pp. Price 50 cents.

The money received from the sale of these publications is deposited in the Treasury, and the Secretary of the Treasury declines to receive bank checks, drafts, or postage stamps; all remittances, therefore, must be by POSTAL NOTE or MONEY ORDER, made payable to the Librarian of the U. S. Geological Survey, or in CURRENCY, for the exact amount. Correspondence relating to the publications of the Survey should be addressed ·

TO THE DIRECTOR OF THE
UNITED STATES GEOLOGICAL SURVEY,
WASHINGTON, D. C.

WASHINGTON, D. C., *November*, 1890.

www.ingramcontent.com/pod-product-compliance
Lightning Source LLC
Chambersburg PA
CBHW021827190326
41518CB00007B/766